Smart Cities in a Digital World

Smart Cities in a Digital World

Editor: Francis Walcott

STATES
ACADEMIC PRESS
www.statesacademicpress.com

Published by States Academic Press
109 South 5th Street,
Brooklyn, NY 11249, USA
www.statesacademicpress.com

Smart Cities in a Digital World
Edited by Francis Walcott

International Standard Book Number: 978-1-63989-483-3 (Hardback)

Cataloging-in-Publication Data

Smart cities in a digital world / edited by Francis Walcott.
p. cm.
Includes bibliographical references and index.
ISBN 978-1-63989-483-3
1. Smart cities. 2. Cities and towns--Computer simulation. 3. City planning--Technological innovations.
4. Information modeling. I. Walcott, Francis.
TD159.4 .S63 2022
307.760 285--dc23

Contents

Preface

A smart city is a framework that aims to develop, deploy and promote sustainable development practices to address the growing challenges of urbanization. It is primarily composed of information and communication technologies (ICT). An intelligent network of connected objects and machines that transmit data using wireless technology and the cloud is essentially a big part of this ICT framework. It consists of cloud-based Internet of things (IoT) applications to help municipalities, enterprises and citizens make better decisions by receiving, analyzing and managing data in real-time. Citizens can pair devices and data with the physical infrastructure and services of the city which helps in cutting costs and improving sustainability. IoT can help communities improve energy distribution, consolidate trash collection, reduce traffic jams and even improve air quality. This book includes some of the vital pieces of work being conducted across the world, on various topics related to smart cities. There has been rapid progress in this field and its applications are finding their way across multiple industries. The book aims to equip students and experts with the advanced topics and upcoming concepts in this area.

The information contained in this book is the result of intensive hard work done by researchers in this field. All due efforts have been made to make this book serve as a complete guiding source for students and researchers. The topics in this book have been comprehensively explained to help readers understand the growing trends in the field.

I would like to thank the entire group of writers who made sincere efforts in this book and my family who supported me in my efforts of working on this book. I take this opportunity to thank all those who have been a guiding force throughout my life.

Editor

1

Learnings from Pilot Implementation of Smart City by a Brazilian Energy Utility

Daniel Picchi, Mateus Lourenço, Alexandre da Silva,
Daniel Nascimento Jr., Eric Saldanha,
Inácio Dantas and José Resende

Abstract

This chapter describes the experience acquired during the implementation of a smart grid pilot project in a Brazilian utility. Learnings in the area of smart metering, telecommunication, information systems and project management are presented. A special focus on Brazilian specificities is given as well as on the management of innovative projects.

Keywords: smart grid, smart metering, advanced metering infrastructure, meter data management, project management

1. Introduction

Smart grid technologies are a set of techniques, systems and equipment that aim to automate grid operations, collect grid information and optimize power distribution, both in power flow and in technical services. These technologies are of great value for energy utilities, since they help reducing costs, avoiding future investments and enhancing its main key process indicators (KPIs). However, benefits coming from the application of smart grid concepts not only may reach the organizational context but also should be transferred to the community in which the companies are present. In that sense, smart grid technologies should be applied in order to result in a smart city, where the energy resources are efficiently used and there is a community awareness about the use and the benefits of such technologies. For that purpose, Elektro has selected the city of São Luiz do Paraitinga, located in São Paulo State, in Brazil, in order to concentrate several smart grid initiatives and evaluate the impacts in the company and on the society, collecting experience to help build a roadmap of future rollout.

In the Brazilian context, the application of some smart grid technologies is in its early stages, with changes needed both in the regulation and the market [1], which is still being developed and some imported solutions may not present prices that result in a positive business case. Some examples are smart metering, whose massive rollout would introduce a representative increase in utility investment and would lead to a tariff increase or a need of government funding, which need further regulatory discussion. Distributed generation is another example that clearly has its financial benefits for the customer, but, due to the current prices, it is still a high investment and long payback application, which is not viable to most people. In this scenario, the massive application of these technologies to create pilot projects is being subsidized by research and development funding from the Brazilian energy agency (ANEEL—Agência Nacional de Energia Elétrica), and a national research has been developed by a group of utilities in order to describe possible scenarios for smart grid deployment [2]. Cybersecurity is also an issue being addressed by ANEEL funded research projects [3], since the introduction of communications on the meters creates a risk of data disruption and also utility infrastructure unsolicited access.

The city of São Luiz do Paraitinga was chosen for this pilot project due to several aspects: the number of customers is representative (5500) and the investment necessary to change all the meters was within the funding limits; the city has extensive rural area and a concentrated urban area, which is a characteristic present in most of Elektro's concession area; the city has a set of hills, which introduce limitations to radio technologies to be studied; the city is a Brazilian historic heritage site, which allows us to understand the relation between the deployment of new equipment and the maintenance of the visual characteristics of the city. The scope of the technologies deployed is below:

- Smart metering: 1.800 ultranarrowband PLC meters + 3.500 RF mesh meters + 80 PRIME PLC meters.
- Distributed generation: 275 solar panels deployed in 12 buildings with generation of 63.25 kW and one wind power generator of 2.4 kW.
- Public lighting: 120 LED remote controlled public lights.
- Distribution automation: 12 automation points, including reclosers, voltage regulators, capacitor bank and fault sensor.
- Electric vehicles: seven electric bicycles and one electric bus.

To engage the customers, a customer portal has been developed in order to present the energy consumption profile and help introduce an energy efficiency culture. The portal is integrated with the meter collection system and can also control the consumption by the means of a consumption goal. The process of meter reading and connect/disconnect has also been fully automated through the development and integration of meter data collectors (MDCs), meter data management (MDM) and customer information system (CIS).

During all the phases necessary to reach these results there has been important lessons learned. The chapter will present in detail the scope of the project in each area of concentration and bring the relevant knowledge acquired during the specification, architecture, deployment and operation stages in the hope of illustrating the challenges of bringing smart grid technologies to real use cases and helping further smart grid deployments.

2. Smart meters

In the project, smart meters have been used in residential customers not only for billing purposes, but also to collect further information of value for the company and the customers. In order to assess the main technologies in use for smart metering, it has been decided to deploy two different kinds of PLC technology, narrowband and ultranarrowband, as well as RF mesh (915 MHz) (**Figure 1**).

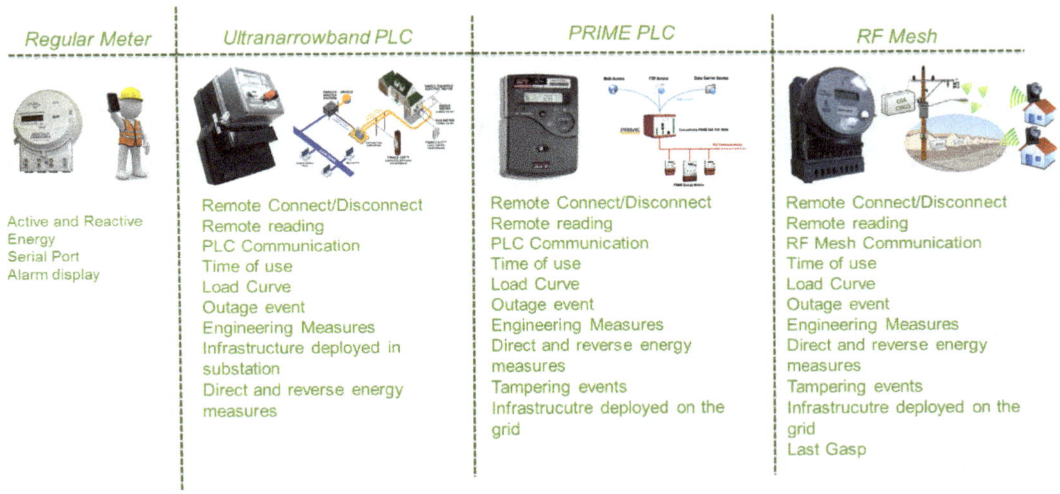

Figure 1. Smart metering technologies.

The process of testing and approving the smart meter models was set in a way that could replicate as closely as possible the field conditions, in special the communication issues.

Currently, Elektro buys usual electronic meters, with a serial output that allows the communication with an external device to collect only meter reads (active and reactive energy). To elaborate a technical specification, a study of the smart meters offered in the Brazilian market has been lead and a specific document and test plan have been developed for the project (**Table 1**).

Note that not all the models on the market covered the whole requirements, therefore, some desirable requirements had to be removed in order to have a sufficient number of meter vendors during the negotiations. However, the requirements removed did not impact the main objectives of the project. This shows how the smart metering market is still under development in Brazil. A lot of vendors have recently developed new meter models to address this new demand and not all of them attend specific utility requirements or have the necessary legal conditions to be marketed. The evolution of the product is a four-hand process in which both the suppliers and the utilities should work closely in order to understand the value of each functionality and shape the products to their reality, which may vary from country to country and even across one single market (**Table 2**).

Requirement	Group	Weight	Compliant models
INMETRO[1] certification	Industry and regulatory standards	19%	77%
Time of use (TOU)			87%
Engineering measures			90%
Clock deviation ≤30 ppm			90%
Metering error ≤1%			100%
Load curve	Basic requirements	27%	77%
Remote connect/disconnect			100%
Optical port			90%
Remote communication			100%
Reconfiguration logs			87%
Remote clock synchronism			70%
Remote firmware upgrade			70%
Overvoltage and undervoltage events	Desirable requirements	54%	70%
Voltage, current and power factor measures			90%
Tamper event—load with no voltage			67%
Last gasp			90%
Reverse energy register			90%
Reactive energy			100%
Duration of reverse flow			20%
Tamper event—voltage on the load side during disconnection			100%
Tamper event—meter opened			60%
Tamper event—reserve energy			90%
Tamper event—meter tilt			80%
Tamper event—DC magnetic field on meter			30%
Duration of meter powered up			10%
Local clock synchronism			40%

[1]INMETRO is the Brazilian Institute of Metrology, legally responsible for approving all the meters available to be marketed.

Table 1. Initial smart meter requirements and adoption.

Requirements	Compliant models (%)
Local clock synchronism	40
Duration of reverse flow	20
Tamper event—meter opened	60
Duration of meter powered up	10
Tamper event—DC magnetic field on meter	30

Table 2. Removed smart meter requirements.

2.1. Testing and validation

Once the meter vendors were filtered by the technical specification, the testing phase took place in Elektro's laboratory, evaluating four samples of meters of each model, summing up 120 samples.

Precision, functional and communication tests were performed. The precision tests were performed according to Brazilian regulation RTM431 [4], from INMETRO. Some of the functional requirements had to be developed during the validation process, demonstrating the innovation characteristic of the project. **Table 3** shows some of the functionalities specially developed for some meter models.

Requirement	Group	Compliant models (%)
Reconfiguration logs	Basic requirements	87
Remote clock synchronism		70
Remote firmware upgrade		70
Overvoltage and undervoltage events	Desirable requirements	70
Tamper event—load with no voltage		67

Table 3. Specific purpose developed functionalitites.

2.2. RF mesh communication tests

For the RF mesh smart meters, a small network was configured in one of the company's premises. The chosen place allowed evaluating the distance between the network elements in which signal attenuation begins to affect communication performance.

The concentrator was located inside a laboratory, while the smart meters were placed in the spots named by "MIB", with an average distance of 70 m between them. As it is shown in **Figures 2** and **3**, the areas in which the concentrator and the MIB1 point are located present a variety of metallic structures and equipment that interfere in radio signal.

MIB2 and MIB3 points were configured to evaluate the behavior of the network for long distances, where a hop between meters is expected. It means that these meters would not communicate directly with the concentrator, but would do it using another meter as a switch.

This behavior was confirmed in the tests and, for some technologies, distances between MIB1 and MIB2 had to be shortened by half in order to establish communication (**Figure 4**).

Figure 2. Area of tests.

Figure 3. MIB1 location.

For the MIB4 point, an enclosed construction was chosen, where a higher attenuation of RF signal was expected and more instability. Instability was measured by the continuous pinging of the meter for the period of test. It was true for most of the technologies, and one specific technology showed no issues with signal attenuation and presented stable communication. Connectivity tests and data collection tests were also successful (**Figure 5**).

Figure 4. MIB2 and MIB3 location.

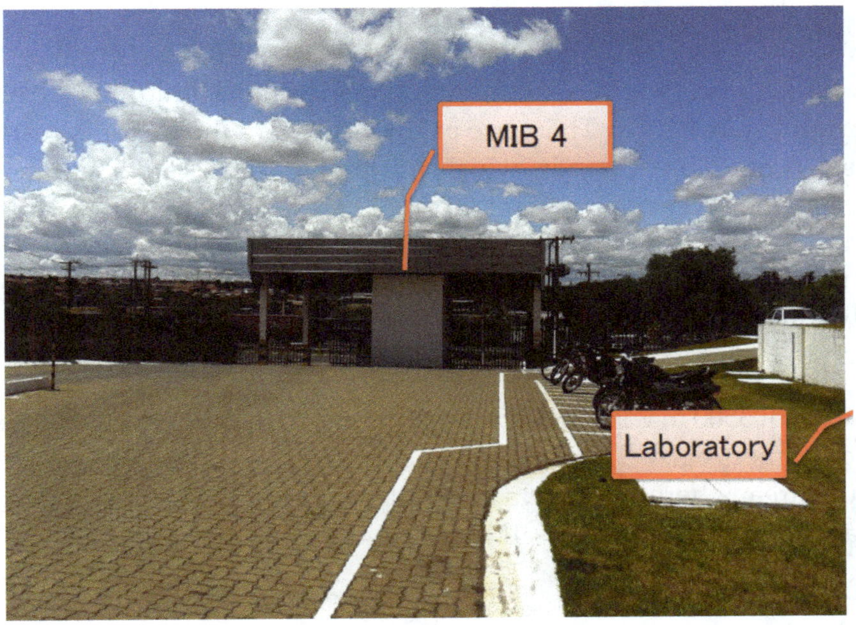

Figure 5. MIB4 location.

3. Telecommunication

The telecommunication network is one of the most important components for the construction of an advanced metering infrastructure (AMI). For the smart city project in Sao Luiz do Paraitinga it has been decided to test three different communication solutions for the last mile and a WiMAX backhaul in unlicensed frequency.

- RF mesh
- Narrowband PLC
- Ultranarrowband PLC

3.1. RF mesh

For communication access to the last mile (NAN—neighborhood area network), it has been chosen a RF MESH IPv6 solution, IEEE 802.15.4e/g, unlicensed frequency, 900 MHz, which ensures a good range when compared to solutions using 2.4 GHz.

The bandwidth for this solution reaches 150 kbps sufficient to traffic information from meters, even when submitted to multiple hops.

The amount of data generated per meter in a residential customer is relatively low, it is not necessary to have high data rates. Energy consumption for transmitters installed on the meters should be low, preventing the increase of technical losses in the distribution system.

To achieve these requirements, the solution was to use the IETF 6LoWPAN—RFC4944—IPv6 [5] over low power WPAN.

The mesh network is the best option for the municipality, since buildings are old and many of them listed by public patrimony. There are examples where the meter is installed in unconventional places. There were cases in which the meter was located within the customer's residence and even behind the refrigerator. In these cases, it would be hardly possible that the meter communicates directly with the data concentrator. It is necessary to use other meters that are closer to the concentrator, creating a more reliable path. To ensure low latency and sufficient bandwidth to traffic information without the need for retransmissions, the limit of hops was limited to 6.

For communication of 3500 m with RF MESH technology it took 20 concentrators with capacity for up to 5000 customers each. The need for 20 equipment was due to low customer density, especially in rural area, where customers are scattered throughout the municipality area. The average occupancy per concentrator is 177 m, and the most occupied one has around 700 m connected to it.

Despite the studies having indicated the need for 20 RF mesh concentrators, not all equipment have been deployed, due to backhaul installation difficulty in the rural area. These points would require investments in expanding the power grid and signal repeating devices, bringing the final communication cost per point as much as 50 times higher than the value of one point of the urban area. Thus, seven concentration points have been canceled, disabling the communication with 694 m.

The RF mesh solution was not feasible technically and financially in areas with low density of meters, requiring high investment in infrastructure to install the backhaul communication and provide connectivity to the AMI concentrator. Another point we should consider is the underutilization of the concentrator, which has the capacity to handle up to 5000 customers and in this environment would not reach even 100 customers.

3.2. Narrowband PLC—PRIME

To test the functionality of the solution in the Brazilian environment, the PoweRline Intelligent Metering Evolution (PRIME) technology of narrowband PLC has been deployed, which enables the transmission of data on the grid with speeds up to 1 Mbps. This solution is widely applied in Europe, presenting excellent results. It has a telecommunication architecture open and nonproprietary focusing on interoperability and scalability. The full interoperability of the solution creates very homogeneous competition between vendors. As a result, PRIME meters are among the cheapest smart meters on market, being a huge advantage to distribution utilities.

Two data concentrators have been deployed to communicate with 75 m in two distinct transformer circuits. The solution presented a satisfactory result, being able to communicate with the meters daily at least six times. This result enables also connect/disconnect commands to be executed by the meters with time intervals from around 4 h, which is a completely reasonable timeframe for such kind of service.

As the data signal is injected into the secondary of the power transformer, one data concentrator for each power transformer is needed. With this operational feature, the solution becomes more financially attractive in the cases where the number of customers connected in the power transformer is high. For this project the average number of customers per transformer is less than 20 customers, which is the reason why the testing of the solution was limited to two data concentrators. However, lower cost data concentrators are being developed and recently appeared on the market, making the solution more attractive to be used in a wide range of grid arrangements.

3.3. Ultranarrowband PLC

To test a smart metering solution for customers located in the rural area of the municipality of Sao Luiz do Paraitinga, an ultranarrowband PLC solution that enables two-way communication on the medium and low voltage distribution grid was installed, being able to exceed transformers on the grid and reach distances greater than 100 km, a common distance in the distribution network in rural areas.

The implemented solution uses the power distribution grid as the path of communication from the meter connected to the low voltage up to medium voltage system at the substation. The solution was installed to cover the four feeders that make up the electrical distribution system of this municipality, thereby, any meter with this technology that is installed in the electrical system is able to communicate with the PLC system.

The solution is composed by a collection software installed at the Data Center, equipment deployed at the substation, such as CTs and processing units, and the meters.

3.3.1. Substation communication equipment

The main components that composed the SCE are:

- Central receiving unit—CRU
- Outbound modulation unit—OMU
- Modulation transformer unit—MTU

- Inbound pick up unit—IPU

Once the CRU receives the command of the central system, the CRU sends the instructions across the electrical network via OMU by MTU for the meter. The meter sends a response to the command also across the power grid and the response is captured by the IPU which forwards the response back to CRU. The CRU responds back to the central system.

The CRU sends out commands to the OMU. The OMU then sends the command to the meter by a slight variation of current in the zero crossing of the 60 Hz sinusoid. This mode of operation ensures the solution the range of dozens of kilometers from the power substation.

Similarly, the communication originated from the meter to the IPU is obtained by sending a signal near the crossover point of zero volts. When the CRU sends commands to the meter, the IPUs await an answer in a particular way over the power grid.

Due to this way of information transmission, the bandwidth is very low, on the order of bits per second, limiting the application to one or two daily readings. On the other hand, it does not use signal modulation at high frequency, therefore, it is possible to transmit data over long distances, not suffering relevant attenuation or signal jamming/filtering by other equipment belonging to the electricity distribution grid (e.g., transformers, capacitors, etc.).

Despite the low data transmission capacity, the solution meets the project requirements. The fact that the municipality has only one power substation and only four feeders facilitated the equipment installation process that made up the solution.

3.4. Backhaul

To provide communication between RF mesh concentrators, DA devices, distributed generation and PLC concentrators, a WiMAX network was built using unlicensed frequency –5800 MHz. In this frequency range, Brazilian legislation limits the transmission power of +30 dBm, which becomes a problem when it is desired to achieve distances over 3 km in a hilly terrain, such the one found in Sao Luiz do Paraitinga.

With these features and limitations, it was necessary to build a new communication network, using point-to-point radios in frequency of 900 MHz. Despite achieving a lower bandwidth comparing to WiMAX radio, up to 1 Mbps, the network meets the needs, mainly to attend AMI concentrators which have few connected meters.

All antennas were installed on poles that serving the power grid, with an average height of 6 m above the ground.

A hilltop near the central area of the municipality was used for the installation of base stations. This hill is at an altitude of 935 m above sea level, while the urban area of the municipality is at an average altitude of 760 m above the sea. This difference in altitude allowed the use of a conventional pole with 11 m in height to the antennas installation, rather than a tower, significantly reducing deployment costs of the backhaul network for the urban area.

Even using this hill, it was necessary to install several repetition points to enable the links to the AMI concentrators installed in the rural area. In the example below, it took two points of repetition, using point-to-point radios at 900 MHz (**Figure 6**).

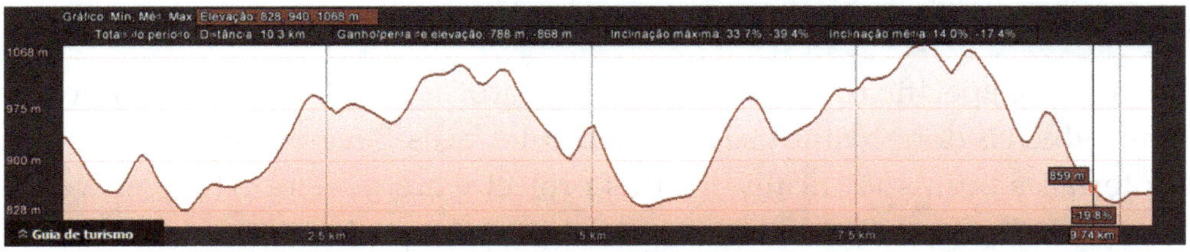

Figure 6. Topology of one link that needed repetition.

For installation of these repeaters, it was necessary to extend the electrical network and install MV/LV transformers to make it possible to feed the radios.

It was not financially feasible to do adequacy in all the necessary points of repetition, being prioritized the points with the highest number of customers per AMI concentrator.

4. Information systems

4.1. System architecture

Data collected and addressed to field devices must be managed by specialist systems. In a general smart metering architecture, the following systems are involved:

- MDC—meter data collector: it collects periodically data from meters and forwards it to the MDM; manages command stacks and is the point of interface between the MDM and the field equipment; implements metering protocols, in order to convert data received and sent to meters to adapt it to the formats expected in interfaces with the MDM; sometimes also plays the role of a network management system (NMS), managing the network in order to access the meters, usually happens when meter reading is not done in transparent mode.
- MDM—meter data management: it manages all metering processes.
- OMS—outage management system: it manages outage information, being able to correlate outage events in order to aggregate them and associate to the most probable point of failure.
- WMS—workforce management system: it manages service orders to be executed by the company specialized human resources and allows operators to schedule and prioritize execution of services.
- NMS—network management system: it manages the communication network equipment, being able to collect communication statistics and control equipment configuration, connectivity and access.

- CIS—customer information system/billing: it manages customer information, such as relation of equipment in customer premises, billing parameters, billing cycles and further commercial processes.
- Customer portal: web interface in which customers access services offered by the company and can visualize information related to their accounts.

All the systems have specific roles in the metering processes, an information flow must exist between them, which is done by the use of system interfaces. Several technologies are available, but all the interfaces built specifically for this project were developed using webservices. In **Figure 7**, the flow of information between systems for the project is presented.

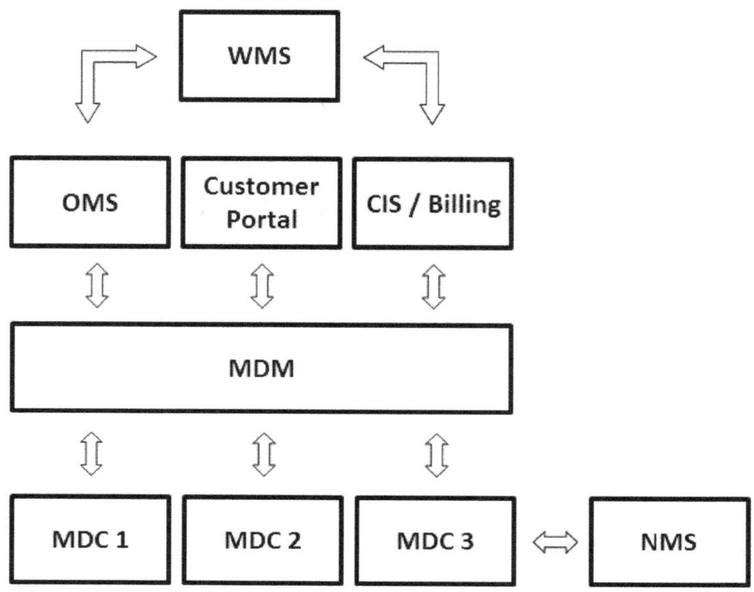

Figure 7. System architecture.

Note that there are three MDC systems, which is not a peculiarity of our implementation. The MDC is intimately related to the meter protocol used on the devices and also the communication technology, therefore it is not unusual to find several MDCs connected to a single MDM. In this project, three different sets of technologies were evaluated, resulting in three different MDCs.

For the RF mesh solution deployed, the communication network is based on IP, and there is a dedicated NMS to manage the radio network, providing connectivity to the meters. In this case, the MDC has to exchange information with the NMS in order to discover the IP addresses of the meters and be able to communicate transparently with them. The NMS also receives last gasp alarms (power outage) from the NIC card present on the meter and forwards it to the MDC, which forwards it to the MDM.

The MDM is the central piece of the solution, gathering information from all the meters, regardless of which MDC they belong to. Its database contains information that must be synchronized with the CIS, such as the meter related to each customer, the billing determinants, the billing cycle and other customer characteristics that are used in the MDM for other processes, such as losses detection. The MDM is also integrated with the OMS, being able to report power outages in its supervised customers. For billing purposes, in Brazilian regulation, only one read is necessary per month, and the MDM sends only the necessary amount of information to the CIS in order to complete the billing process. However, the meters contain much more detailed information, such as hourly consumption, which is of great interest to customers. For that purpose, the MDM is also integrated with a web-based customer portal, in which customers can follow their load profile and also establish a consumption goal, which is monitored by the MDM, which send notices when the customer has reached a predetermined threshold or when the projected consumption exceeds the customer goal.

To complete the processes, some actions on the field must be executed, mainly reading meters that failed the billing process and re-establishing power outages. These actions are executed through service orders, which are generated by the CIS and OMS and managed by the WMS.

4.2. MDC/MDM interfaces

Except for MDC systems, all other components of the architecture are unique and can have specific crafted interfaces. In the case of MDC/MDM interfaces, however, it is more effective to have one single specification, which reduces further costs of integration of the MDM with new MDC systems that might be deployed. This is possible, since the main services that a MDC must provide to the MDM are the same for all the meter and communication technologies. There will be some specific implementations, but the broader scope of the integration is to be less expensive and quicker to put in production.

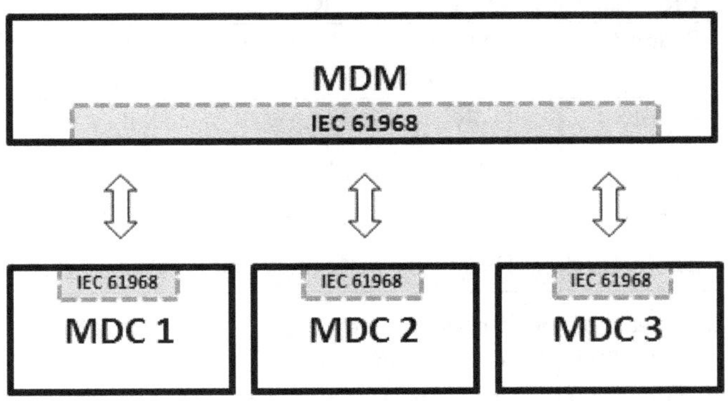

Figure 8. IEC 61968 connectors.

For the project, the services provided by both systems were defined following the IEC 61968 standard [6], which defines a framework of XML structures to represent the most common needs of utility system integrations. A single connector on the MDM is capable of integrating with different instances of the IEC 61968 interface (**Figure 8**).

The interface covers methods for synchronization, reading, configuration and events reporting. **Figure 9** is an example of the message flow for an event query made by the MDM onto the MDC.

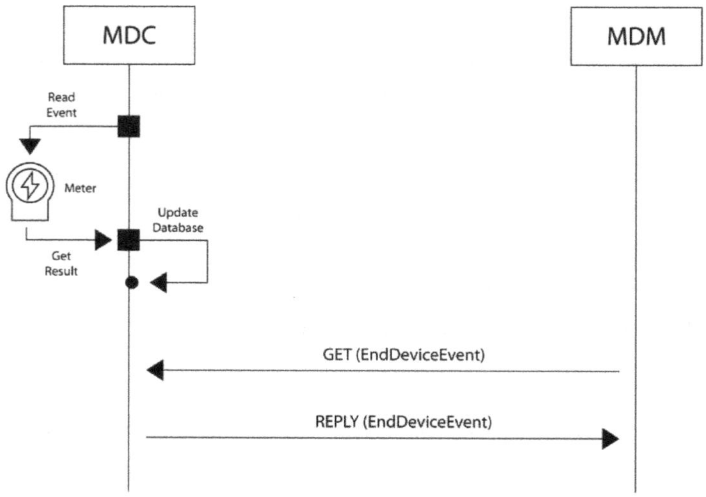

Figure 9. Message flow for one IEC 61968 method.

4.3. The value and cost of data

The MDM is a data repository that can contribute to several areas inside a utility, from energy demand planning to asset management. All the process of collecting and organizing this data is the most important role of the MDM and feeds all its other functionalities. The system is loaded with templates for each type of equipment, defining the variables to be read and the frequency and resolution of each reading type.

Reading	Ultra-narrowband PLC		PRIME PLC		RF mesh	
	Resolution	Reading frequency	Resolution	Reading frequency	Resolution	Reading frequency
Accumulative registers[1]	Hourly	Daily	Hourly	Daily	Hourly	Daily
Load curve[2]	NA	NA	Hourly	Daily	Hourly	Daily
Engineering measures[3]	Instant	Daily	Instant	Daily	Instant	Daily
Events	On demand	Daily	On demand	Daily	On demand	Daily

[1]kWh and kVArh per time of use (peak, off-peak, shoulder, etc.).
[2]Array of kWh and kVArh information with hourly timestamps.
[3]Instant values of current, voltage, frequency, power (W, VA and VAr).

Table 4. Reading frequency and resolution for smart meter's data.

It is very important to understand that there is a limitation in quantity and frequency of readings per meter, which is determined by the communication technology, the meter hardware and the processing performance of the systems and interfaces (**Table 4**). In the project, we have initially worked with the following configuration:

The treatment given to each reading is very important to assure the most cost-effective solution. On one hand, bringing too few information might not suit the business needs; on the other hand, bringing too much information might make information access slower and increase storage costs. Therefore, the amount of readings of each variable must be tuned in order to be meaningful and just sufficient.

For instance, RF mesh and PRIME PLC technologies would allow 15-min readings of energy consumption. But the average consumption of the customers where the meters were deployed is 0.07 kWh for this time interval. There is no meaning in collecting this amount of data, since it is more likely to be redundant information. Hourly reads are sufficient for this use case.

On the opposite way, instant readings of voltage and current collected only once a day by ultranarrowband PLC have not contributed to any application of this information. Since there is no meaning in it, these readings could be disabled from the reading process on the MDM, saving several gigabytes of useless data.

4.4. Smart metering processes

The main smart metering processes that the MDM enables are:

- Remote meter reading
- Billing
- Nontechnical losses detection
- Load profile monitoring
- Outage detection

As mentioned in the last topic, the remote meter reading process feeds information to all the other processes.

Billing uses the registers collected daily to calculate the billing determinants and send the information to the CIS in the dates corresponding to each billing cycle. Having a periodic reading allows the system to make reading estimates in the case which the information is not collected on the days preceding the billing of a customer, reducing the impact in the number of service orders sent to the field.

Comparing the consumption from a transformer and all the meters connected to it, we are able to calculate the losses for each secondary circuit. The system then compares this value to a defined threshold and notifies an operator when probable nontechnical losses are found.

The same consumption can be also monitored by the system and compared to a goal defined by the customer through the web portal. When a certain threshold is reached or when the consumption projection by the end of cycle exceeds the goal, the customer is then notified by e-mail or SMS.

Some meters support the last gasp functionality, in which, under a supply loss, the meter sends an alarm before shutting down. This alarm is then forwarded throughout the systems and is received on the OMS that manages the outage and can correlate the information with the information received from other meters, allowing a much more accurate fault detection.

5. Project management

Once this is not a project management book, it will not describe generic tools to implement project management. The project was managed following the market best practices [7] and in this section it will be transposed a little part of experiences, difficulties and lessons learned about how to manage a smart grid pilot project. It may be possible to apply these tips to another type of project design, however it is necessary to consider that the knowledge of this section was acquired on a smart grid application developed in Brazil.

The first important experience to share is how to build a technical specification. This is a very important part of the project; in this step, we can determine the success or failure of the project. This activity is the key process to maintain your project in line with your project management plan. Usually, to write a technical specification, the project manager focuses only on the technical characteristics of products, leaving aside some points that will be fundamental to the quality and project costs. Some important questions that must be answered in a technical specification are related below, each of these items might result in delays or changes in project schedule.

- Which technologies, equipment and technical features are expected?
- Which technical validations are necessary for the equipment and systems approval?
- Where each component of the solution is produced? Is it produced on a third party factory? Is it necessary to import some components?
- Is there subcontracting? In which terms?
- Which training or certification is required?
- How will the project schedule be? This item can be subdivided into several parts, below are three important points:
- Which are the contract terms: technical rounds, trade rounds, internal approvals, contract signatures, production factories, imports, deployments, tests, bug fixes.
- Which problems can impact deadlines: delays in subcontracting, problems in product development, imports, shortage in skilled labor, pioneering deployments.
- The political and economic scenario can influence directly on acquisitions, productions and deadlines, exchange rate changes, tax increases and reduction of subsidies.
- How will the payments to all suppliers be? It is an important point since several problems may occur during the project. Hold, block or suspend payments, supported by the contract, can be effective management tools.
- How interest clauses and penalties will be designed?

Project management means dealing with the unexpected, depending on the extent of the impacts, legal action should be taken. In technical specification it is necessary to establish how the penalties will be applied, this is a tool that could help in harsh situations. It is noteworthy that the application of contractual penalties should be applied in a manner that does not compromise the structure of the project or any supplier. Evaluate project points, deliveries or developments that are on the critical path of schedule, divide the macro project into several subprojects with milestones, deliverables and well-defined responsibilities, with penalties endorsing it, use this as a way to mitigate future risks.

So, how can we map all these points before signing the contract? It is recommended that the procurement process can be divided into parts, which can be carefully evaluated before signing the contract.

5.1. Procurement process

5.1.1. Request for information (RFI)

This is the preliminary technical specification which describes all questions related to equipment, financial health of the company and experiences that must be considered to choose a partner. Through this instrument an initial assessment of suppliers in the market, their prices, their products and relevance on the subject can be made.

5.1.2. Technical visit

During the visits, it will be possible to know the company structure. Possible doubts and solutions can be addressed in order to get the necessary support to have a very complete technical specification and with low risk exposure.

5.1.3. Technical meeting with suppliers

The project technical team needs to discuss with suppliers every issue related to technical implementations. This is the moment when the project team will define or change the technology assumptions related to the expected solutions.

5.1.4. Site survey

The site survey is the moment when the suppliers check the technical assumptions to evaluate and measure the project implementation risks. This step is essential because an error in that process can result in either over or undersized project costs; in both cases this could derail the project.

5.1.5. Preparation of technical specifications

The technical specifications should be established after the knowledge of the previous stages, which technologies to be applied, which solutions were verified in the field, which are the risks and which actions will be taken to mitigate them. At this point we should be aware on aspects of the schedule, delivery and payment method for this type of solution. The payment will be made after commissioning or upon delivery of each equipment?

5.1.6. Request for proposal (RFP)

Submission of all specifications defined with the market. It is noteworthy that the amount of technical specifications is directly linked to the contracting strategy and contract management, a fact that must be decided in the strategy of the project.

5.1.7. Response to RFP

Moment in which suppliers respond to the technical specification by presenting their solution proposal in technical and commercial terms.

5.1.8. Technical evaluation

Formal response to the technical specifications, selecting which suppliers meet all the technical requirements.

5.1.9. Commercial negotiation

Moment in which the commercial negotiation is made among all technically validated suppliers to define the most suited proposal to the project's budget. It is important to notice that sometimes the cheapest solution is not the most suitable, since it may present higher maintenance cost or adaptation needs.

From that moment, we are able to select a set of suppliers that are capable to meet all established requirements in the technical specification process. The next stage is how to elaborate the contract, due to the importance of this activity and the relevance of its impacts, the next session will be devoted to address some of its particularities. This is a topic of great matter, since smart grid is a new concept in Brazilian market and many companies are developing brand new products to supply utility needs. Since these are the very first experiences both for the utilities and for the suppliers, extra care is needed in writing the contract to avoid any issue that could stop the project due to operational or financial impacts.

5.2. Contract management

Once we have built a good technical specification, the next critical process is the contract management. It is very important to rely on tools that can ensure that the supplier can match the project requirements. A good financial analysis is the first step, after this it is important to check who are the subcontractors and who are the suppliers of your provider. China is relevant player to manufacture components or equipment assembly. Exchange rate changes can dramatically affect your budget, for this reason it is essential to define the maximum exchange rate risk, insert in your contracts tools to control and monitor exchange variation. You will need to alert your supplier of this risk, a planning failure of your supplier might lead to delays in your schedule or even new disbursements in your project.

Once you are secure with the financial issues it is time to focus on the technical requirements. To help in these issues it is important to answer some questions: Is the technology expertise of your supplier sufficient to provide the solutions needed to your project? Which are the finished products and which are the products that need to be developed? The technical specification helps to answer these questions, however, in this section we need to use this information to design the penalties. To use it as a scope control tool, each relevant milestone should be related to a penalty. Penalties should be sized in order to represent the importance of the milestone, but not high enough to impact financial health of the supply. It is important to have penalties that, once triggered, have their values raised over the time in which the supplier is not attending its compromises. A contract should not be designed to harm its partners. Suppliers are partners in the project, so the penalties must be used only in critical situations. Never place all penalties on a single item or delivery, apply it in partial and time-spaced deliveries. If your supplier delays, try to establish a dialog before the imposition of fines, a conversation is always the best solution, however a small and constant fee can help change the direction of the project and helps engaging all the resources towards the project accomplishment.

5.3. Scope controlling

This section will describe important tips for monitoring project activities and how to monitor the progress of the schedule; this is the point to deal with the unexpected and the possible changes. A relevant tool might help in this part of the project, it is important to bear in mind how the project scope is performing compared to the scope baseline. It is necessary to monitor the work performance, project management plan updates, project documentation updates and other important updates. Every single update must be planned and monitored in project schedule; a small change in a critical path could be the difference between the success or failure of the project.

Not always your supplier knows how to implement project management skills and that would be an unpleasant surprise. It might be necessary a strong management structure in your supplier, these measures must be implemented in the partners or in the project team to maintain the performance closest to the project baseline. First of all, it is important to identify what is delayed and which is critical, after this you must bear in mind that in most cases it is possible to find an alternative path, a tough action of the project manager can often help to find solutions. It is usual that project partners have many other important developments; it is very relevant to find a way to make aware the partner on the project priorities. This can be achieved by project meetings, penalties or rewards linked to the progress of the project schedule. The project manager must understand the difficulties to help the project team to address problems and monitor the solutions implemented and their impact. It is very desirable that the project team understands the impact of each proposed amendment; it will help the team to design better alternatives. For this, everyone must know the needs of stakeholders and what is expected of each phase or goal to be achieved. It is important that the project leader clarifies to the team the relation between each delivery and stakeholder needs.

It is imperative that the project team and partners trust the project manager, who should always keep a respectful and professional close relationship with each member of the project. The relationship between the project team and the project manager is another critical factor in the success of every implementation. The project leader must seek the same relationship conditions with their business partners; the ideal is that it has free access to all information from their suppliers that can avoid unexpected surprises and provide time to react if any unforeseen happens.

It is undesirable, but in some cases the project manager will need to realign expectations with stakeholders; this can occur for several reasons: the failure of the project team analysis, external conditions changes to the project team or any other event that may have altered the initial conditions. Before realigning expectations, the project manager must ensure that everything that can be done to correct the problem is done, should raise all impacts and delays, should propose new actions that can mitigate risks and reduce impacts. The project sponsor must have all analysis before using his influence to change the project conditions. The external regulatory body should be aware of changes to the project schedule, once the financial resources of a smart grid project are often from regulated funds it is very important that the regulator sees you as a partner with whom you can communicate changes and conditions, to mitigate the disallowance risks or other regulatory issues.

5.4. Project closure

At every stage of the project, it is important to keep a very accurate documentation control; this will be very valuable during project closure phase. In this section it shall be prepared delivery reports of products and milestones reached, the balance sheet and any changes that have occurred in the scope of the project. If the project manager played his role there will be no surprise at this stage, all alignments have already been made, and it will be required only a formalization of each item.

The client must already know all the features and the operation mode of each delivery, however, it is extremely necessary to provide documentation required for the use of each solution implemented. The client must receive technical reports containing the test results of each product or solution, in this manner there should be no doubts about deliveries and their use.

All the tools used during the cost management and all generated documentation will be of great help to the financial closing, the disbursement control linked with the project baseline will help it to explain the costs of project, every change must be documented to become part of the final financial report.

All research carried out must have its reports and characteristics found; we should be able to clearly see certain requirements as applicability and relevance to the activities of the project. In the case of a project that uses regulatory funds for research and development, criteria such as originality and reasonability of costs must also be considered.

All information must be present in a final report, which must contain clearly each of the requirements, the manner in which they have been met, the costs and how all research and learning were managed and led to the final product.

Author details

Daniel Picchi*, Mateus Lourenço, Alexandre da Silva, Daniel Nascimento Jr., Eric Saldanha, Inácio Dantas and José Resende

*Address all correspondence to: daniel.picchi@elektro.com.br

Elektro Eletricidade e Serviços S.A. (Elektro Electricity Services), Campinas, Brazil

References

[1] Redes elétricas começam a ser modernizadas no Brasil (Electric grids start to be modernized in Brazil) [Internet]. 2015. Available from: http://especial2015.canalenergia.com.br/redes-inteligentes [accessed: 2016-28-01].

[2] Kagan N., et al. Redes elétricas inteligentes no Brasil: análise de custos e benefícios de um plano nacional de implantação (Smart Grids in Brazil: national deployment plan cost benefit analysis). 1st ed. Rio de Janeiro: Synergia; 2013. 300 p. ISBN: 978-85-61325-87-9.

[3] Suiama D., et al. Avaliação de segurança em medidores eletrônicos (Smart meters cybersecurity assessment). 1st ed. Salto: Schoba; 2015. 352 p. ISBN: 978-85-8013-427-8.

[4] Portaria INMETRO/MDIC number 431, INMETRO Std. RTAC 431, 2007.

[5] Transmission of IPv6 Packets over IEEE 802.15.4 Networks, RFC Std. 4944, 2007.

[6] Common Information Model (CIM)/Distribution Management, IEC Std. 61968, Part 9, 2009.

[7] Indelicato G. A guide to the project management body of knowledge (PMBOK® guide), 4th ed. Proj Mgmt Jrnl; 2009. 40:104. doi: 10.1002/pmj.20125.

2

Smart Brain Interaction Systems for Office Access and Control in Smart City Context

Ghada Al-Hudhud

Abstract

Over the past decade, the term "smart cities" has been worldwide priority for city planning by governments. Planning smart cities implies identifying key drivers for transforming into more convenient, comfortable, and safer life. This requires equipping the cities with appropriate smart technologies and infrastructure. Smart infrastructure is a key component in planning smart cities: smart places, transportation, health and education systems. Smart offices present the concept of workplaces that respond to user's needs and allow less commitment to routine tasks. Smart offices solutions enable employees to change status of the surrounding environment upon the change of user's preferences using the changes in the user's biometrics measures. Meanwhile, smart office access and control through brain signals is quite recent concept. Hence, smart offices provide access and services availability at each moment using smart personal identification (PI) interfaces that responds only to the personal thoughts/preferences issued by the office employee not any other person. Hence, authentication and control systems could benefit from the biometrics. Yet these systems are facing efficiency and accessibility challenges in terms of unimodality. This chapter addresses those problems and proposes a prototype for multimodal biometric person identification control system for smart office access and control as a solution.

Keywords: office access using brain signals authentication, office appliances control, brain signals capture, analysis and interpretation

1. Introduction

Building smart office systems implies building a system that recognizes an employee and interacts with employees through reading their brain signals and interpreting their brain activities and signals patterns to control their offices. The control involves controlling the light brightness off/low intensity/high intensity and temperature increase/decrease, chair height or back angle, curtains up/down, and doors lock/unlock status. Concerning the infrastructure required for building smart offices, it is worthy as it enhances accommodating important category of people with disabilities and improves their employability. For ordinary people, it will provide flexible working environment by adding comfort and fun to the workspace [1].

The smart office perceives intentions and responds to their intended needs by actuating the environment [2]. Hence, designing smart offices involves sensory data from reading their brain signals, temperatures, etc., and hence auto responds to their need in terms of controlling office light brightness off/on and temperature increase/decrease, chair height or back angle, and curtains up/down status. This employee-office interaction will save them some time and will increase the work efficiency and effectiveness as well as adding a strong helpful tool to those who have struggles doing such a thing. For ordinary people, it will also add some fun and make offices happy zones by acquiring employee's thought signals, and they will have a flexible working environment. This smart human-office interaction requires up-to-date sensory devices as well as capturing devices to be used for collecting employee's intentions and hence send it to interpretation system for identifying the commands to be executed for access or controlling particular office item.

2. Literature review

Designing smart offices and smart environment has been reported in the recent research as biometric technologies that deploy human-computer interaction. These technologies fall into either the personal identification or command-based systems. Personal identification systems imply identity recognition such as finger print, eye print, voice print, and palm vein data for personal identification systems. Command-based systems use eye gaze, voice commands, etc. In addition, a most recent research has reported the use of more advanced interaction levels such as the use of brain signals and emotion recognition systems for both personal identification and controlling office devices.

An attempt to develop an intelligent emotion stress recognition system (ESR) using brain signals (EEG) is published in the field of biomedical engineering and sciences for diagnosis of verbal communication problems and treatment of disability in speech and bodies. Eye tracking was also used by disabled to communicate with the outside world [3]. This research investigates the possibility of how to recognize employee's stress emotions using signal processing of electroencéphalographie. ESR suggested new system recognition for émotionnel stress, using multimodal bio-signals using electroencephalogram (EEG) as the main signals, since its use is spread widely in clinical diagnosis and biomedical research. A cognitive model is then used to extract the brain signals from the appropriate EEG channels that represent emotional stress relevant data [4].

Generally speaking, any EEG-based system would go through the following units: signals capturing unit, signals preprocessing/processing unit, classifier, and decision-making unit that translates the classified signal, **Figure 1**.

2.1. Current EEG applications

One of the leading projects in building smart environment was a Smart Environment for offices at University of Stuttgart (Sens-R-Us) application [5]. Sens-R-Us project focused on using graphical interface (GUI) and Mica2 motes sensors that capture the real-world data of their employees. These sensors are static sensors and personal sensors. The base sensors are installed in all rooms such as office and meeting rooms, and they send location beacons with room ID constantly. Personal sensors are carried around by the employees, and they receive location beacons and then select the highest signal base stations. Personal sensors can also send signals to update their information which is used in a constant detection of meeting occurrence. The developed Sens-R-Us application advantages are the use of lower power consumption, small size, possibility to be used off offices. The GUI is used to acquire employee's information and status, room temperature, and available switched on devices in the office. Another application reported by the literature is WSU "Smart Home." WSU is an assistive technology that aims to assist elderly people in performing daily routine tasks home. The "smart home in a box" is about 30 sensors application which detects motion, temperature, and power sensors that are easy to install. The system provides functionalities such as monitoring and learning the elderly routines, recording changes when arise and remind elderly if they forgot something [6].

2.2. EEG authentication applications

Over the past decade, unimodal authentication systems occupied the top place in many fields, for example, finger prints in students/employees attendance system, eye print/face

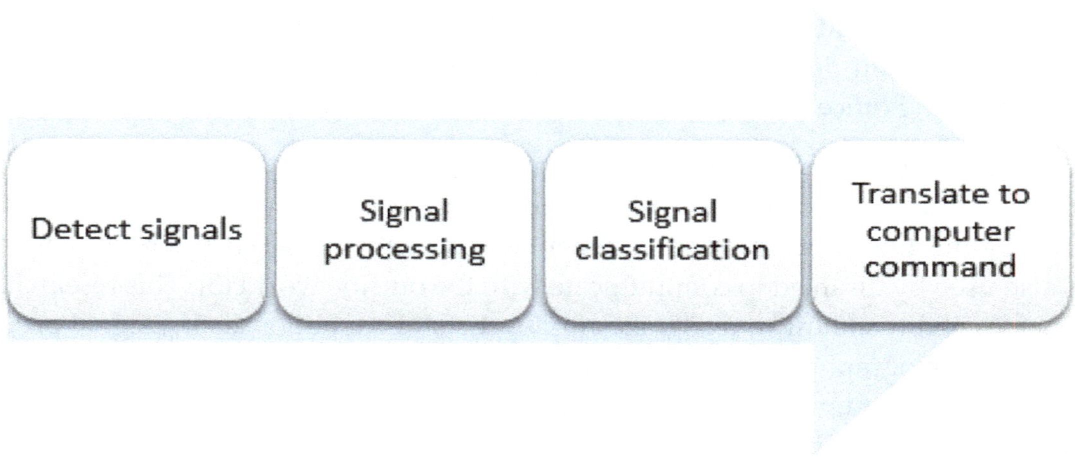

Figure 1. BCI processes.

recognition in the airports, etc. Most of the demonstrated problems in unimodal biometric systems are noisy data such as scars in the skin of fingerprints, defects in the capturing sensor, limited number of degrees of freedom that result in feature similarities with large population, recording the voice and use it to get access to voice recognition systems. Also, they justified using multimodal biometric system as a solution since the main cause of efficiency problems imposed by unimodal biometric systems is the reliance on the evidence of one source of information.

Ross and Jain [7] introduced smart office access system based on multimodal biometric technologies. They developed multimodal biometric system taking into consideration appropriate fusion for output of different modalities, strategies to integrate the models [7]. There are some examples for the development of biometric multimodal systems such as face recognition and fingerprint multimodal biometric authentication system by Rahal et al. [8]; Face recognition and speech-based multimodal biometric authentication system by Soltane et al. [9], Al-Hudhud et al. [10]; and speech, signature, and handwriting features authentication system by Eshwarappa [11]. Other research is concerned with a new biometric model known as electroencephalography or EEG, and it is a type of wave signals produced by the brain, mostly used in applications related to brain health/research. Researchers have suggested that the EEG has a potential as a powerful authentication model since some features that are extracted from the EEG signals are unique from one person to another [12–15]. A survey conducted by Khalifa et al. [12] presents several methods used in EEG authentication, please refer to **Table 1**.

Unlike "what the user have" authentication such as iris and fingerprint or "what the user knows" such as password variants of authentication, Mohanchandra et al. [16] introduced a "what the user is" authentication type as an application that uses real-time EEG signals for locking/unlocking the computer screen. It matches mental task encoded features (MTEF) of the EEG through Euclidean distance measurement with MTEF of current EEG user status. The result has been shown that the system is a reliable system of authentication. Additionally, the results presented a good classification accuracy that, however, needs some improvements [16].

Building EEG-based mobile biometric authentication systems was initiated by Klonovs and Petersen [17]. They proposed a system that uses EEG and NFC tags. Users would choose their

Technique	Channels	Users	Task	TAR	FAR
A	2	40	Rest	79%	21%
B	6	4	Rest, math, letter, count, rotation	–	0.1% average combination using five features
C	–	8	Rest	80%	
D	15	9	Left/right hand movement	95% (left)94.81% (right)	

Table 1. Khalifa et al. [12] presented the following classification accuracy rate when deploying EEG in authentication including task measure in terms of true acceptance rate (TAR) and false acceptance rate (FAR).

own personal password as an image in the enrolment phase; EEG data will be obtained from the headset from four EEG sensor locations: P7, P8, O1, and O2. In the access time, this image would be shown to the user when authenticating in a five seconds period of time. The authors chose zero crossing rate technique [26, 27] and wavelet analysis [34] for feature classification and latencies measurement of visual-evoked potentials, respectively. Potentials, respectively. The result of their work is that they have found that the most significant features can be extracted from the visual parietal-occipital cortex of the brain, and thus, their implementation of presenting the image method can be seen beneficial.

2.3. Signals capturing types

In BCI, there are three different methods to get the signals: invasive, partially invasive, and noninvasive. In the invasive method, the BCIs are implemented directly into the gray matter of the brain (usually used to help paralyzed people). However, although this method gives high-quality signals, it still presents some risk to human health. On the other hand, in the partially invasive BCIs, only part of the BCI is implemented inside the skull but not within the brain. However, the signal captured has lower resolution than the invasive method. In addition, it presents lower health risk on the patient. For the noninvasive method, sensors are placed on the scalp and no implanting needed. This method does not present any risk to human health, and it is convenient and easy to use. Additionally, it provides good signal readings [18, 19].

2.4. Signals acquiring techniques

There are different methods to obtain brain signals. One of them is by electrical means such as electroencephalogram (EEG) where sensors called electrodes are used to acquire the signals. This method has low set up cost and ease of use. However, it is susceptible to noise and requires intensive training before using it. Other methods to acquire the signals are by nonelectric means such as measuring by the magnetic and metabolic changes or even from the pupil size oscillation as developed lately in Ref. [20], all these techniques can be used in a noninvasive manner. The functional magnetic resonance imaging (fMRI) technique uses magnetic to capture the brain activity, and it focuses on measuring the blood oxygenation and flow that is increased in the area of the brain which involves mental activity. Therefore, it requires large devices with a large magnetic field scanner. The functional near-infrared spectroscopy (fNIRS) uses infrared waves to measure the blood oxygenation and flow. However, most of the techniques that depend on measuring the metabolic changes suffer from long latency compared to the EEG technique [18, 19].

2.5. BCI device types

The term headset is used to describe the capture devices and may include shapes of cap, tiara, headband, helmet, or even loose electrodes. Many commercial headsets have been released to the market with an attractive design and low cost. NeuroSky [21, 22] and Emotive EPOC [23] are examples of these devices. Most of these applications improve their performance by using brain signals as an input along with other parameters such as body temperature or pupil size, and it also combines the BCI technology with other technologies such as the virtual reality [18, 19].

2.6. Thought identification

There are different ways to identify thoughts or mental activities that resulting action such as motor imagery, bio/neurofeedback for passive BCI designs, and visual-evoked potential (VEP). In motor imagery, imagining moving any parts of the body results in sensorimotor cortex activation, which modulates sensorimotor oscillations in the EEG [24]. The second type is the bio/neurofeedback for passive BCI designs, where the relaxation and attention were measured using some bodily parameters along with mental concentration that is measured by monitoring the alpha and beta waves of the brain [18, 24]. On the other hand, the VEP captures the brain response to visual stimulus such as certain flashing graphic elements or sound stimulus such as special sound pattern [18, 24].

2.7. Feedback type

Zander and Kothe [35] introduced the categories of BCI approaches as follows:

- Active BCI: independent of external events, useful for controlling an application [25].
- Reactive BCI: arising in reaction to external stimulation. It is indirectly modulated by the user for controlling an application [25, 33].
- Passive BCI: It refers to the brain activities that are integrated to produce an input. The integrated input applies mental state BCI in which the user does not try to control his brain activity [23].

In addition, there are other classifications for BCI approach that depend on the processing and rhythm types. There are two types of BCI processing: **online** which happens while the user utilizes the BCI and **offline** which occurs after experiment. On the other hand, the rhythm's classification is divided into two types: **synchronous** where the commands are processed after every certain amount of time and **asynchronous** where processing the commands will be upon the request [18].

2.8. Application area

The development of the BCI technology makes it no longer used only at laboratories but anywhere (home, offices, etc.) since it became a portable device. Therefore, its applications are also growing and becoming more different and advanced to many areas such as communication and control, motor substitution, entertainment, motor recovery, and mental state monitoring [19].

2.9. Group of BCI Beneficiaries

The basic group of BCI beneficiaries is the disabled patients to help them in their remembering & managing daily tasks and expressing themselves. In addition, BCI has been used in rehabilitation of disorders such as stroke, addiction, autism, ADHD, and emotional disorders [19].

Dues to the development of new applications in fields other than medical field, new groups are emerging recently as BCI technologies main users groups. Among these new emerging fields are authentications & security systems, health applications, controlling games, biometrics, and controlling of smart and virtual environments [19].

3. Proposed technical solution

The proposed solution for accessing and controlling smart offices system includes two main modules.

First module tackles a multimodal biometric accessibility system that includes electroencephalography (EEG) and face recognition part in addition to a nonbiometric part, known as SMS token. This part describes the feature extraction from the cloud storage of the biometric data and the best multimodal fusion technique for the biometric and nonbiometric combination [31, 32].

The other module is the smart office control. This module includes controlling office devices through brain signals. The office devices control would be highly demanded in very busy schedule for workers in terms of saving time to walk away from the desk in order to increase/decrease the light brightness or the temperature of the office. In addition, it is important to embed infrastructure for cases of people who have major disabilities that prevent easy movements and actions. A mental control for smart workspace module is introduced in this chapter that is based on acquiring brain signals that represent the workers thoughts regarding their feelings of temperature and lighting in their offices. Hence, the brain signals are processed and filtered [28, 29] in order to analyze and interpret the feeling in terms of frustration and willing to increase/decrease any of the surrounding status. Mental control system would require a smart working environment that is equipped with sensors and actuators. All these components together with the data collected from brain signals are used to anticipate any need.

3.1. The proposed authentication subsystem process units

The core functionalities for the proposed system implementation and overall processing are as follows:

(a) Data acquisition unit: EEG signals through an EEG headset and face images through web camera.

(b) Signal & Image preprocessing and filtering unit: for ALL signals and images being acquired.

(c) Feature extraction unit for each model, brain signal, and face images.

(d) Face recognition unit.

(e) Multimodal fusion unit all on real time.

(f) Classification and decision-making unit.

3.1.1. *Data acquisition unit*

Brain signal data are acquired during both enrolment and login phases for each biometric modality. The system administrator defines authorized users and provides them with passwords. Once an authorized user enters his name, the EEG enrolment phase starts by displaying the images representing the password. The user chooses the password image from a photo gallery and then confirms. Hence, the EEG signal capturing instructions is displayed for the user which indicates that the signal will be captured for 5 seconds. During that period, the chosen image will appear and the user should focus on it without blinking and in a relaxed condition for 5 seconds.

The user will repeat the same procedure done in the enrolment phase without the first step (choosing the password image). The following steps take place at the login phase:

(a) The user enters his name.

(b) The user is forwarded to the EEG log in page.

(c) Brain signals will be captured only twice; the total time of each recording is 5 seconds.

(d) The system will select the brain signals channels AF3, F3, and F7 located at standard positions of the international 10–20 system.

(e) The raw data representing the captured signals are then written to a CSV file.

(f) The other modality capturing; face recognition, starts. The system will forward the user to face image capturing page, and instructions appear to inform the user to look at the camera.

3.1.2. *Preprocessing and filtering*

The captured EEG signals are written in CSV, named the raw data. Raw data are noisy, that is, it contains a lot of irrelevant information. A preprocessing step is needed to extract the relevant features. Spatial EEG data are prepared by zero amplification to a power of two in order to be Fourier transform to frequency domain. The next step was to remove the baseline activity from each channel and then calculate the mean of each channel. Hence, the mean of each channel was subtracted from the original values of the channel. The transformed data is filtered with a 5th order sinc filter and band pass filter with frequency range between 0.5 Hz and 60 Hz to notch out 50 Hz and 60 Hz. The sampling rate is 128 Hz. Hence, the system calculates the mean of each channel, and then, the mean of each channel is subtracted from each original value of the channels. Finally, we applied the inverse Fourier transform.

3.1.3. *Feature extraction unit*

The system initiates EEG feature vectors that are saved temporarily in the runtime memory during the enrolment phase, to extract the following features:

- Signals speed (frequency),
- Power spectral density,
- Magnitude, signal-to-noise ratio,
- Variance, mean, and standard deviation for each channel,
- Zero crossing rate from each channel.

The features vector is then stored using either local server or a cloud storage. Hence, the following processes and calculations take place:

(a) Means of all channels are calculated and chosen as the baseline.

(b) The variance is calculated for each original channel signal values.

(c) The low pass filter is applied to reject frequency higher than 40 Hz.

(d) The data are then filtered and processed by removing the baseline activity from each channel.

(e) The extracted features are standard deviations (SD) from each channel. The training pattern was five from each subject, and the SD average was calculated and saved as stored features. This was implemented by the authors previously and published [10, 30, 31, 32].

3.1.4. *Face recognition modality*

During enrolment and login phases, the system prompts the face recognition modality interface that captures the face image. The face image is then converted into gray scale. Haar Cascade classifier is applied to the grayscale image for face recognition. Cropping and resizing (to 1010 × 100) step is done to the face images. The resulted image is encrypted with AS encryption technique and lastly saved in the file system as a training set. By this the will be fully enrolled in the system.

At the login phase, the applies the same steps being performed during the enrolment procedure. The system applies face recognition using principal component analysis (PCA). First, the system will:

1. Retrieve the enrolment features vector.
2. Decrypt the enrolment features vector using AS algorithm.
3. Calculate combination of some component or face basis called the "Eigenface" from the features vector.
4. Face space is calculated by projecting the face images.
5. Euclidean distance between the detected face and each image is calculated from "Eigenface."

6. Recognize if the distance is above the distance threshold. The system temporarily saves the face recognition matching decision.

3.1.5. Classification and decision-making unit

The classification is performed using Euclidean distance, the classifier would perform in a way that the interclass (distance between the groups) was maximized, and the intraclass (distance within the same group) was minimized. The Euclidean distances for three patterns were computed for each subject to result in three thresholds for that subject. Both thresholds are saved for each subject in the enrolment phase. The classification is performed for the brain signals and for the face image using the following classifiers:

1. *Cosine similarity*

$$\text{Similarity} = \cos(\theta) = \frac{A.B}{AB} = \frac{\sum_{i=1}^{n} A_i x B_i}{\sqrt{\sum_{i=1}^{n} (A_i)^2 x \sum_{i=1}^{n} (B_i)^2}}$$

2. *Euclidean similarity*

$$d(p,q) = d(q,p) = \sqrt{(q_1 - p_1)^2 + (q_2 - p_2)^2 + \cdots + (q_n - n)^2} = \sqrt{\sum_{i=0}^{n} (q_i - p_i)^2}.$$

The classifier computes the threshold for each subject for the input modality by comparing the enrolment and login feature vectors. The average threshold is computed from the five enrolment trials and saved as stored threshold. Authentication is done during the login phase using five patterns from the subject, and the new resulted threshold is averaged from the five patterns. Then, the new average threshold is subtracted from the stored threshold.

Personal identification decision score is produced in the matching process for each input modality. The decision can be either: **Accept** the subject if the classifiers average thresholds from two classifiers are less than 0.100, or **Reject** the subject if the classifiers average thresholds from two classifiers are less than 0.100.

3.1.6. Fusion unit

In this stage, the system gets the decision scores of both modalities the brain signals (EEG) and the face recognition. The system then will:

1. Reject access if ALL decision scores are "*Reject*"
2. Grant access if ALL decision scores are "*Accept.*"

However, if the system gets one **Accept** from one modality with **Reject** from another modality, then the system uses the SMS token. The SMS token works by sending a system-generated one-use password in the form of SMS to the registered mobile number. The is then

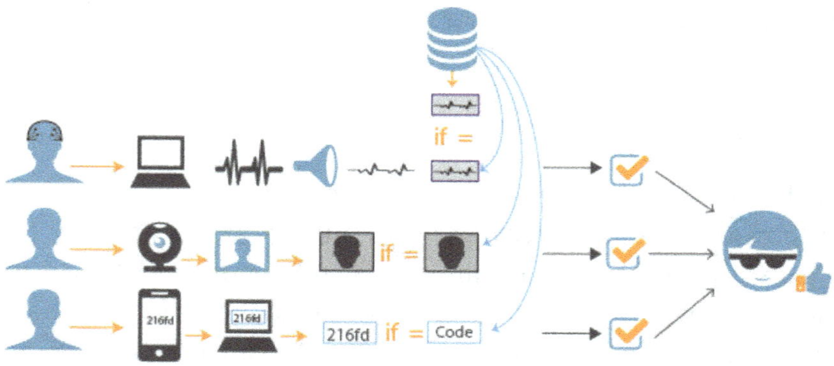

Figure 2. Decision scores for both modalities and fusion process.

redirected to a page where the received password can be entered password to verify the. If the password is correct then the grant access and if not access will be denied; please refer to **Figure 2**.

3.2. Controlling the office devices using EEG signals

This section describes the subsystem that is used for smart offices control through recording the brain signals during the brain activity when thinking of increasing/decreasing the temperature and increasing/decreasing the light intensity. For each activity, brain sensory data are passed to the system so that it can be encoded into a command. The system stores these commands in the form of vector feature associated with the. This subsystem integrates the following: a simulator which will be designed using 3D modeling tools to model the offices, sensors, and devices to be controlled through the brain thoughts, Emotive Headset to read brain signals, and then interfacing tools to integrate and produce the interface (see **Figure 3**).

In order to build a smart office that allowing employees to control their offices temperature and brightness, this subsystem will integrate physical devices, brain signals coming from the Emotive Headset, and computing entities in offices with the interfacing tools needed to

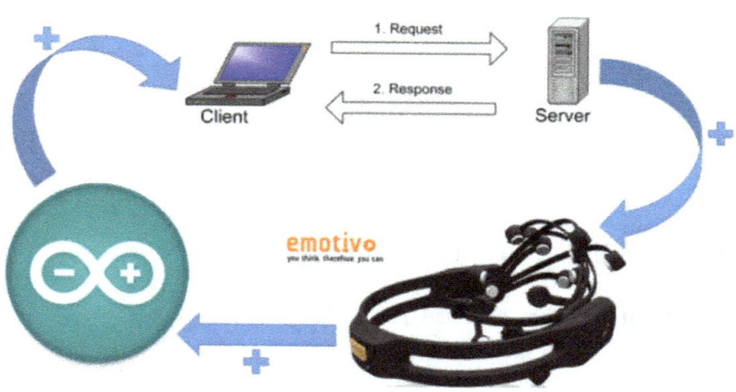

Figure 3. Controlling physical devices by brain signals concept diagram.

produce the interface. In addition, interpretation of the thoughts will be translated into a command that will be passed to the actuator of the specified device, **Table 2**.

Hardware	Software
The 3D models in the offices for the physical appliances (sensors, fan, and light bulb) Arduino set	Software to control these physical appliances
	Computing entities: Preprocessing units Filtering unit Classification unit Decision-making unit
Emotive headset	Brain signals capturing unit Brain signals analysis Recognition unit
	A scheduler for planning and action execution
	Interface

Table 2. Office control subsystem components, please, refer to **Figure 2**.

4. Experimentation setup and results

A total of 30 people volunteered to participate in our study. The subject was instructed to put on the Emotive EPOC EEG headset and is asked to follow these steps:

1. The subject will be seated on a normal chair, relaxed arms resting on their legs, and in noise-controlled room.
2. The will be exposed to the GUI of the system.
3. The chooses an image from an image gallery.
4. The mental task that is to focus on the particular image of a celebrity for 5 s, during which the signals will be captured; a task during which the should concentrate and not moving the body nor blinking.
5. The brain signals are recorded and forwarded to the next step.
6. The subject will be looking at the camera in order to allow the system to capture the face image.
7. The system then will process both input modalities: EEG and face image in order to produce the feature vectors.
8. The features vector is then compared to the stored feature vector for each participant, and experimentation results are presented in **Table 3**.

The experiments are designed such that the user will be wearing an Emotive EPOC EEG headset and will be provided with instructions for completing the session. The first instruction

Iteration number	Channels	Task	Users	FAR (%)	TAR (%)
1	P7,P8,O1,O2	Visualizing	7	18	42
2	AF3,F7,F8	Visualizing	3	43	80
3	AF3,F7,F8	Visualizing	3	26	100
4	AF3,F7,F8	Visualizing	32	21	74
5	AF3,F7,F8	Visualizing	32	14	88

Table 3. System performance and summery experimentation results.

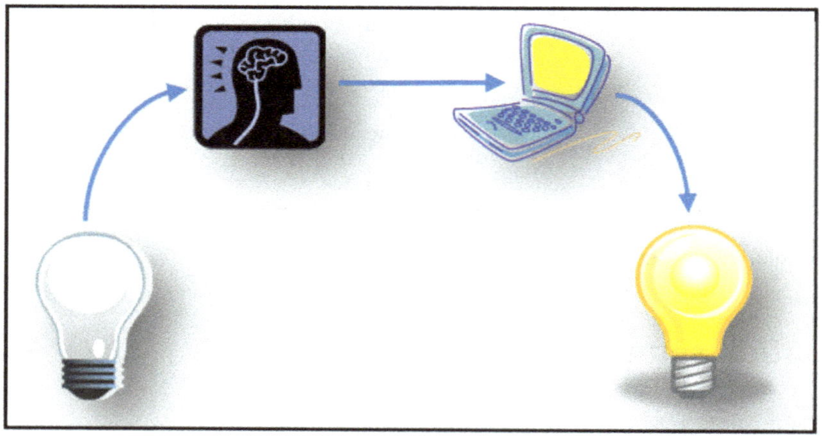

Figure 4. Brain signal control for smart office light intensity.

Figure 5. Physical prototype with Arduino kit for light intensity changes with brain commands.

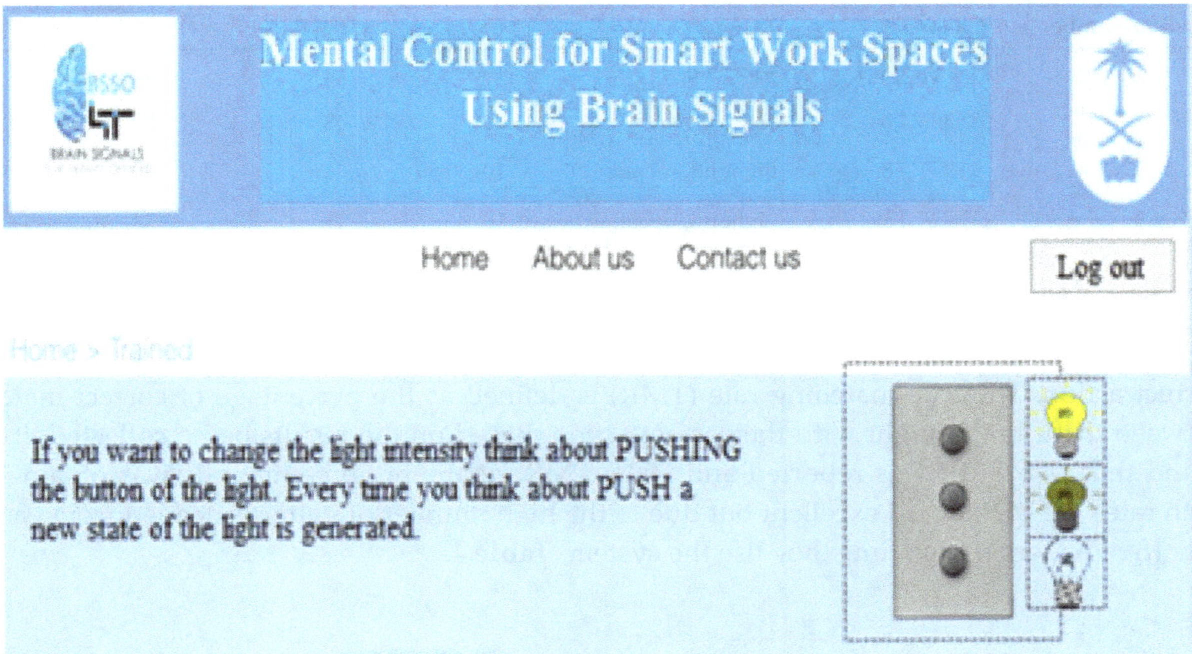

Figure 6. Interpretation of the brain signals in terms of changing the light intensity interface.

Figure 7. Physical prototype with Arduino kit embedded in the office prototype for light intensity changes with brain commands.

will be to ask to do a mental task that is to focus on increasing the temperature for 6 s, during which the signals will be captured. The EEG data were recorded, filtered, and processed as the same way described in the previous section (see **Figures 4–7**).

The results are interpreted in terms of the false matching rate and true matching rate. False matching rate (FMR) is defined as the percentage of matching false user's thought with the

Iteration number	Channels	Task	Users	FMR (%)	TMR (%)
1	P7,P8,O1,O2	Temp up	10	11	42
2	AF3,F7,F8	Temp down	10	43	80
3	AF3,F7,F8	Light int. increase	10	26	90
4	AF3,F7,F8	Light int. decrease	10	14	74

Table 4. Summary of iteration performance regarding the true matching rate (TMR) and false matching rate (FMR).

correct action. The true matching rate (TMR) is defined as the percentage of correct match between the users thought with the correct action. Based on the results being collected, it is found that 26% FMR was reported and 100% TMR for the brain commands was obtained. Both rates are considered excellent but due to the high number of patterns needed from each user (five patterns) each time they use the system, **Table 4**.

5. Conclusion

The work presented here investigated two main terms: first, brain signal and what is the perfect way to read the signal and translate it into real action and second, smart offices and its use in real time.

The work focuses primarily on smart access to the office and smart control of the office devices. Hence, a model was proposed in the chapter for reading the thought in the form of brain signal, translating the thought into password for accessing the system and hence creating other control actions in the office. This requires many sensors in the work environment to receive the translated action and apply it.

Regarding the smart accessibility of the office, the work investigated the use of three authentication modalities as a multimodal authentication system to overcome the limitations of unimodal biometric authentication systems. However, according to the experimentation results, the multimodal system has proven to overcome the efficiency, accessibly problems, fusion mechanism for the multiple models, and the immaturity of EEG model in the field of biometric authentication.

Regarding the smart office access, the model focuses primarily on using EEG as an authentication biometric and secondly, on face recognition. In addition, the proposed solution also investigates the multimodal fusion technique that combines all system models (electroencephalography, face recognition, and SMS token). The authors also referred to reported research results to decide on the most suitable channels from the extracted brain signals using the EEG and multimodal.

The major contributions done through this work are the findings of the best features, classifiers, and methods that are suitable for EEG in authentication and control. For this kind of multimodal system, the findings have been shown the best fusion level to present a

Criteria	Sens-R-Us	BSSO
Goal	Collecting info from employees in an office	Changing the office state
Way of collecting data	Sensors (static and portable), PC	Emotive headsets
Kind of data collected	Position, room temperature, status	Brain signals
Action	Update database info	Change the office status
Support of people with disabilities	Does not provide extra comfort	Provides extra comfort and shortcuts

Table 5. Comparison between Sens-R-Us [5] and BSSO [32].

powerful and efficient multimodal authentication system with accuracy rates of TAR = 90% and FAR = 0%.

The future improvements suggested are as follows: (a) flexibility regarding the EEG signals acquiring device, (b) improving the classifier and thresholding technique to count for the different concentration levels for the same user, and (c) achieving more accuracy in terms of TAR and FAR.

In conclusion, this prototype opens the gate wide in front of new era of internet of things toward a smarter society needs and requirements. Hence, the research could be the milestone for newer inventions and researches and a helpful contribution in the great field of brain computing interaction for authentication systems.

Table 5 shows a comparison between Sens-R-Us and the brain signal smart office control functionalities.

Acknowledgements

This research project was supported by a grant from the "Research Centre of the Female Scientific and Medical Colleges," Deanship of Scientific Research, King Saud University.
Parts of this chapter are reproduced from authors' recent Human Computer Interaction Conference HCI 2015 publication [32] "Brain Signal for Smart Offices." Springer is the publisher for all the HCI2015 conference papers in the theme of Distributed, Ambient, and Pervasive Interactions as a book chapter, Vol. 9189 (2015), of the series Lecture Notes in Computer Science pp 131–140 Springer.

Author details

Ghada Al-Hudhud

Address all correspondence to: galhudhud@ksu.edu.sa

Department of Information Technology, King Saud University, Riyadh, Saudi Arabia

References

[1] P. MiFkulecký. Smart environments for smart learning. In 9th international scientific conference on distance learning in applied informatics, Štúrovo, Slovakia; 2-4 May 2012.

[2] J. Martin, C. Le Gal, A. Lux and J. Crowley, Smart office: design of an intelligent environment. Intelligent Systems, IEEE, vol. 16, no. 4, pp. 60–66, 2001.

[3] L. C. Sarmiento, P. Lorenzana, C. J. Cortes, W. J. Arcos, J. A. Bacca, A. Tovar, "Brain computer interface (BCI) with EEG signals for automatic vowel recognition based on articulation mode", *Biosignals and Biorobotics Conference (2014): Biosignals and Robotics for Better and Safer Living (BRC) 5th ISSNIP-IEEE*, pp. 1-4, 2014, May.

[4] S. A. Hosseini and M. A. Khalilzadeh, Emotional stress recognition system using EEG and psychophysiological signals: using new labelling process of EEG signals in emotional stress state. In 2010 international conference on biomedical engineering and computer science, Wuhan, 2010, pp. 1–6. doi:10.1109/ICBECS.2010.5462520

[5] D. Minder, P. J. Marron, A. Lachenmann and K. Rothermel, Experimental construction of a meeting model for smart office environments. In Proceedings of the first workshop on real-world wireless sensor networks (REALWSN 2005, SICS Technical Report T2005:09), June 2005, Stockholm, Sweden

[6] Available: http://wsucasas.wordpress.com/1313/06/21/smart-homes-feature/ [Accessed on March 2016].

[7] A. Ross and A. K. Jain, Multimodal biometrics: an overview. In European signal processing conference, Vienna, Austria, 2004, pp. 1221–1224.

[8] S. M. Rahal, H. A. Aboalsamah and K. N. Muteb, Multimodal biometric authentication system—MBAS. Information and Communication Technologies. ICTTA '06. 2nd, vol. 1, pp. 1026, 1030, 2006.

[9] Soltane, Mohamed, Noureddine Doghmane, and Noureddine Guersi. "Face and speech based multi-modal biometric authentication." *International Journal of Advanced Science and Technology* 21.6 (2010): 41–56.

[10] G. Al-Hudhud, N. Alrajhi, N. Alonaizy, A. Al-Mahmoud, L. Almazrou and D. Bin Muribah.Brain signal for smart offices. Book chapter in distributed, ambient, and pervasive interactions, vol. 9189 of the series Lecture Notes in Computer Science, pp. 131–140. Springer. 2015.

[11] M. N. Eshwarappa, Multimodal biometric person authentication system using speech signature and handwriting features. International Journal of Advanced Computer Science and Applications, 2011.

[12] W. Khalifa, A. Salem, M. Roushdy, K. Revett. A survey of EEG based user authentication schemes. InInformatics and Systems (INFOS), 2012 8th International Conference on 2012 May 14 (pp. BIO-55). IEEE.

[13] S. Marcel and J. D. R. Millan, Person authentication using brainwaves (EEG) and maximum a posteriori model adaptation. IEEE Transactions on Pattern Analysis and Machine Intelligence, vol. 29, pp. 743–752, 2007.

[14] I. Nakanishi, S. Baba, & C. Miyamoto. EEG based biometric authentication using new spectral features. In *Intelligent Signal Processing and Communication Systems, 2009. ISPACS 2009. International Symposium on* (pp. 651–654). (2009, January) IEEE.

[15] K. Mohanchandra et al., Using brain waves as new biometric feature for authenticating. International Journal of Biometrics and Bioinformatics, vol. 7, pp. 49–57, 2013.

[16] K. Mohanchandra, G. M. Lingaraju, P. Kambli, V. Krishnamurthy. Using brain waves as new biometric feature for authenticating a computer user in real-time. International Journal of Biometrics and Bioinformatics (IJBB) 7, no. 1 2013: 49.

[17] J. Klonovs and C. Petersen, Development of a mobile EEG-based feature extraction and classification system for biometric authentication. Aalborg University Copenhagen, Copenhagen, Rep. June 8, 2012.

[18] A. L. S. Ferreira, L. C. de Miranda, E. E. C. de Miranda and S. G. Sakamoto, A survey of interactive systems based on brain-computer interfaces. SBC Journal of Interactive Systems, vol. 4, no. 1, pp. 3–13, 2013.

[19] R. Leeb and J. D. R. Millán, Introduction to devices, applications and s: towards practical BCIs based on shared control techniques. In towards practical brain-computer interfaces: bridging the gap from research to real-world applications, pp. 107–129, Biological and Medical Physics, Biomedical Engineering, 2012.

[20] S. Mathôt, J. B. Melmi, L. van der Linden, S. Van der Stigchel, The mind-writing pupil: a human-computer interface based on decoding of covert attention through pupillometry. Public Library of Science One, vol. 11, no. 2, p. e0148805, 2016. doi:10.1371/journal.pone.0148805

[21] "EEG - ECG - Biosensors", Neurosky.com, 2016. [Online]. Available: http://neurosky.com/. [Accessed: 16-Mar-2016].

[22] "Epoc", Emotiv.com, 2016. [Online]. Available: https://emotiv.com/epoc.php. [Accessed: 16-Mar-2016].

[23] L. George and A. Lécuyer, An overview of research on "passive" brain-computer interfaces for implicit human-computer interaction. In International conference on applied bionics and biomechanics ICABB 2010—Workshop W1 "Brain-Computer Interfacing and Virtual Reality", Oct 2010, Venise, Italy. 2010.

[24] C. Mühl, B. Allison, A. Nijholt and G. Chanel, A survey of affective brain computer interfaces: principles, state-of-the-art, and challenges. Brain-Computer Interfaces, vol. 1, no. 2, pp. 66–84, 2014.

[25] M. Poel, F. Nijboer, E. L. van den Broek, S. Fairclough and A. Nijholt, Brain computer interfaces as intelligent sensors for enhancing human-computer interaction. In

Proceedings 14th ACM international conference on multimodal interaction (ICMI'12), 22–26 Oct 2012, Santa Monica, CA, pp. 379–382, 2012.

[26] P. He, M. Kahle, G. Wilson and C. Russell, Removal of ocular artifacts from EEG: a comparison of adaptive filtering method and regression method using simulated data. In Conf. Proc. IEEE Eng. Med. Biol. Soc, vol. 2, pp. 1110–1113, 2005.

[27] P. Senthil Kumar, R. Arumuganathan and C. Vimal, An adaptive method to remove ocular artifacts from EEG signals using Wavelet transform. Journal of Applied Science Research, vol. 5, pp. 741–745, 2009.

[28] N. Mourad, J. P. Reilly, H. de Bruin, G. Hasey and D. MacCrimmon, A simple and fast algorithm for automatic suppression of high-amplitude artifacts in EEG data. IEEE International Conference on Acoustics, Speech and Signal Processing, vol. 1, pp. I393–I396, 2007.

[29] A. Garcés Correa, E. Laciar, H. D. Patiño and M. E. Valentinuzzi. Artifact removal from EEG signals using adaptive filters in cascade. In *Journal of Physics: Conference Series*, vol. 90, no. 1, p. 012081. IOP Publishing, 2007.

[30] G. Al-Hudhud. Affective command-based control system integrating brain signals in commands control systems. International ISI Journal of Computers in Human Behavior, vol. 30, pp. 535–541, 2014.

[31] G. Al-Hudhud, E. Alarfaj, A. Alaskar, S. Alqahtani, B. Almshari, H. Almshari. Multimodal biometric authentication web-based application. In Proceedings of IEEE the 5th national symposium on information technology: towards new smart world that will be held in Riyadh, Saudi Arabia, February, 17–19, 2015

[32] G. Al-Hudhud, M. Abdulaziz Al zamel, E. Alattas, A. Alwabil. Using brain signals patterns for biometric identity verification systems. ISI International Journal of Computers in Human Behavior, vol. 31, pp. 224–229, 2014.

[33] N. Al-Ghamdi, G. Al-Hudhud, M. Alzamel and A. Al-Wabil, Trials and tribulations of BCI control applications. In Proceedings of the science and information conference IEEE explore, London, UK, October 2013

[34] I. Güler and I. G. Übeyli, Adaptive neuro-fuzzy inference system for classification of EEG signals using wavelet coefficients. Journal of Neuroscience Methods, vol. 148, pp. 113–121, 2005.

[35] Zander T.O., Kothe C., Welke S., Roetting M. Utilizing Secondary Input from Passive Brain-Computer Interfaces for Enhancing Human-Machine Interaction In Hofmann A. (Ed.): Lecture Notes in Computer Science, Springer, Berlin Heidelberg, 2009.

3

Wind Farm Connected to a Distribution Network

Benchagra Mohamed

Abstract

This chapter presents power flow study for distribution network connected to wind farm based on induction generators (IG). It provides an overview of wind energy conversion systems (WECS) and their related technologies. The details of turbine components, system configurations, and control schemes are analyzed. Wind farm–distribution network systems are developed by MATLAB/SIMULINK to perform different tests under various operation conditions. The impact of wind speed fluctuation on power flow, grid voltage dynamic stability, and frequency responses are also investigated. The proposed simulator is applied to a 3 MW wind farm, and then the efficacy of the proposed simulator has been validated.

Keywords: renewable energy, wind farm, power control, distribution network, smart grid, power flow, voltage stability, frequency stability

1. Introduction

According to the updated grid codes, wind farms tend to be considered as power generation plants, which should perform in a similar manner to conventional power-generation plants [1]. When large renewable energy sources are integrated into a distribution network, the dynamics and the operations of the network are affected [2]. Installed wind power capacity has been progressively growing over the last two decades. The increase in wind turbine size implies more power output since the energy captured is a function of the square of the rotor radius [3]. The majority of wind turbines operating in the field are grid-connected, and the power generated is directly uploaded to the grid.

The emergence of the smart grid will pose wind energy developers with a new challenge, such as production during high wind speed; wind turbine should be able to continuously supply the grid, and power factor correction (PFC). For wind power applications, wind turbines were mostly operated with induction generators (IGs) [4]. As a network interface, the use of

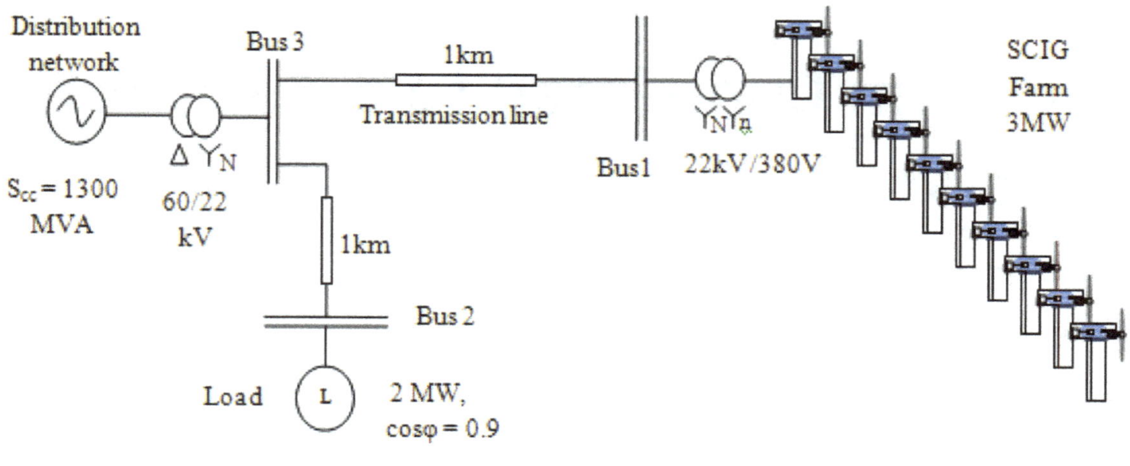

Figure 1. Wind farm–connected distribution network.

converters is extensively reported, showing their capability to extract maximum power in a wide range of wind conditions, and can control both active and reactive powers independently.

The interconnection of wind energy to the power grid is still on the rise. The increasing size of Wind Park resulted in new interconnection grid codes. The power output of wind turbine systems is directly affected by wind speed. Then, when large wind energy sources are integrated into a distribution network as shown in **Figure 1**, the dynamics and the operations of the network are affected [5].

The real and reactive power sharing can be achieved by controlling two independent quantities: the frequency and the voltage magnitude [6]. Any voltage or frequency fluctuation in the wind farm side has direct impact on the grid side [7]. The rotor voltage source converter (VSC) is controlled to maximize the generated power, and the grid VSC is controlled to produce good power quality [8].

The rest of the chapter is organized as follows: in Section 2, the fundamental properties of a WT and its mathematical model are summarized; Section 3 describes the design and control of the Induction Generator and back-to-back converter; Section 4 describes the distribution network connected to wind farm; Section 5 is devoted to the performance simulation of power flow and voltage dynamic stability by MATLAB-SimPowerSystems; and, finally, Section 6 includes the conclusions.

2. Modeling of wind turbine

2.1. Variable-speed wind turbine system configuration

The performance of the wind energy system can be greatly enhanced with the use of a full-capacity power converter. **Figure 2** shows the schematic diagram of the studied wind turbine based on induction generators. All of the subsystem components: WT, VSC, Voltage Source Inverter(VSI), PI controller, transformer, filter, and distribution system, are modeled individually.

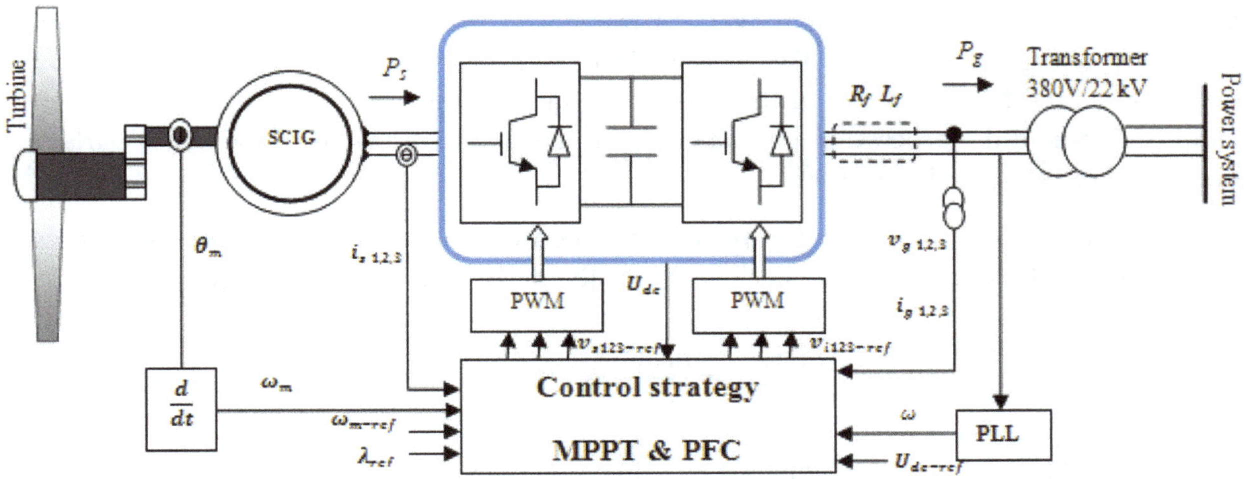

Figure 2. Control of wind turbine.

With the use of the power converter, the generator is fully decoupled from the grid, and can operate in full speed range. This also enables the system to perform reactive power compensation and smooth the grid connection. The turbine rotor shaft of the studied squirrel cage induction generator (SCIG) is coupled to the hub of the turbine through a gearbox for transforming the low rotational speed of the turbine to the required higher rotational speed of the generator. **Figure 3** shows a wind turbine model. The generator converts mechanical energy to electrical energy. The converter is the interface between the generator and the distribution network [4]. The controllers provide proper switching signals for the VSC-SGIC side so as to extract the maximum power from the WT and VSI-grid side to control the voltage and frequency by using a phase-locked loop (PLL) process.

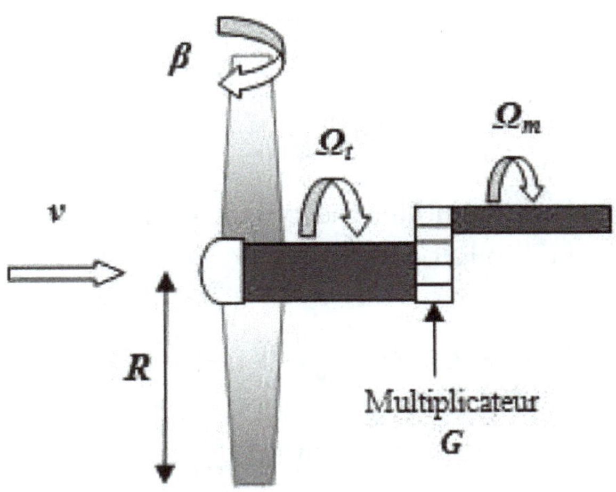

Figure 3. Wind turbine.

2.2. The wind turbine model

A wind energy conversion system (WECS) transforms wind kinetic energy to mechanical energy by using rotor blades. This energy is then transformed into electric energy by a generator, so the turbine is one of the most important elements in wind turbine. In order to better understand the process of wind energy conversion, descriptions of the major parts of a wind turbine are given in this section.

The model of the WT and pitch control is developed on the basis of the steady-state power characteristics of the turbine. The mechanical torque produced by the WT is given by [4]:

$$T_m = \frac{\rho \pi R^2}{\Omega_t} v^3 C_p(\beta, \lambda) \tag{1}$$

where ρ kg/m^3 is the air density, R m is the radius of the turbine, and v m/s is the wind velocity. $C_p(\beta, \lambda)$ is the power coefficient that is dependent on the turbine design, Ω_t is the turbine speed, it is a function of mechanical speed ω_m and gear ratio G, $\Omega_t = \frac{\omega_m}{G}$. β is the rotor-blade-pitch angle, and $\lambda = \frac{\Omega_t R}{v}$ is the tip speed ratio [9]. The gearbox conversion ratio, also known as the gear ratio, is designed to match the high-speed generator with the low-speed turbine blades.

As can be observed from Eq. (1), there are three possibilities for increasing the power captured by a wind turbine: the wind speed v, the power coefficient Cp, and the sweep area $A = \pi R^2$.

The power characteristics of a wind turbine are defined by the power curve, which relates the mechanical power of the turbine to the wind speed. It is a highly nonlinear function. The power curve is a wind turbine's certificate of performance that is guaranteed by the manufacturer [1].

The wind turbine input power usually is

$$P_v = \frac{1}{2} \rho A v^3 \tag{2}$$

The output mechanical power of wind turbine is

$$P_m = C_p P_v \tag{3}$$

We consider a generic equation to model the power coefficient C_p, based on the modeling turbine characteristics described in Ref. [10]:

$$\begin{aligned} C_p(\lambda, \beta) &= 0.5109\left(\frac{116}{\lambda_i} - 0.4\beta - 5\right) exp\left(\frac{-21}{\gamma}\right) \\ \gamma &= \left(\frac{1}{(\lambda + 0.08\beta)} - \frac{0.035}{(\beta^3 + 1)}\right)^{-1} \end{aligned} \tag{4}$$

where λ is defined as the ratio of the tip speed of the turbine blades to wind speed:

$$\lambda = \frac{R\Omega_t}{v} = \frac{R\omega_m}{Gv} \tag{5}$$

Ω_t is the turbine speed, it is a function of mechanical speed ω_m and gear ratio G [11].

Mechanical equation for the drive train (gearbox) is done by

$$J\frac{d\Omega_m}{dt} = \sum T \tag{6}$$

and

$$\sum T = T_m - T_{em} \tag{7}$$

where T_{em} is the electromagnetic torque, and T_m is the mechanical torque. J is the total inertia, it can be calculated as

$$J = \frac{J_t}{G^2} + J_g \tag{8}$$

where J_t and J_g are respectively, the turbine inertia and the generator inertia.

The power captured by the wind turbine is a cubic function of wind velocity (see Eq. (2)) until the wind speed reaches its rated value. To deliver captured energy to the distribution network at different wind velocities, the induction generator should be properly controlled according to the wind speed in order to extract the maximum power available.

Because of the maximum $C_p(\lambda, \beta)$ is obtained at a optimal tip speed ratio $\lambda = \lambda_{opt}$, looking for taking the maximum power available at any wind speed, the control system must adjust the mechanical speed to operate at λ_{opt}.

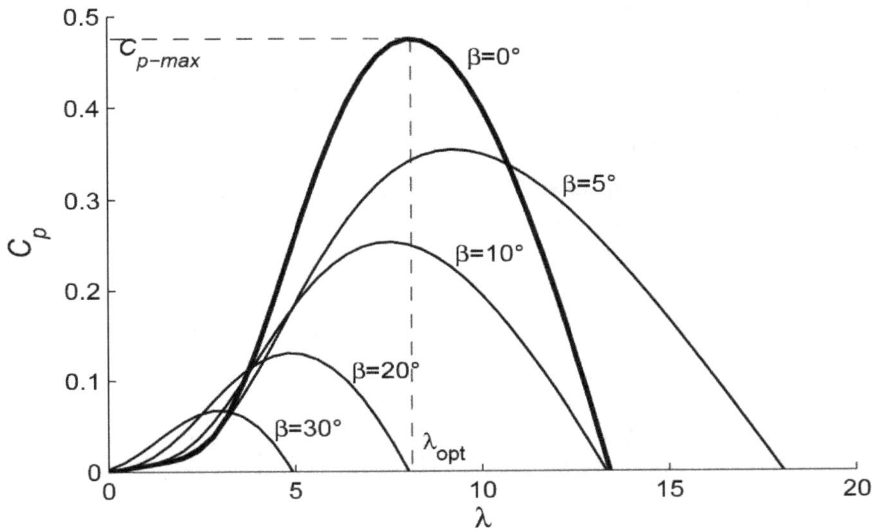

Figure 4. Power coefficient C_p.

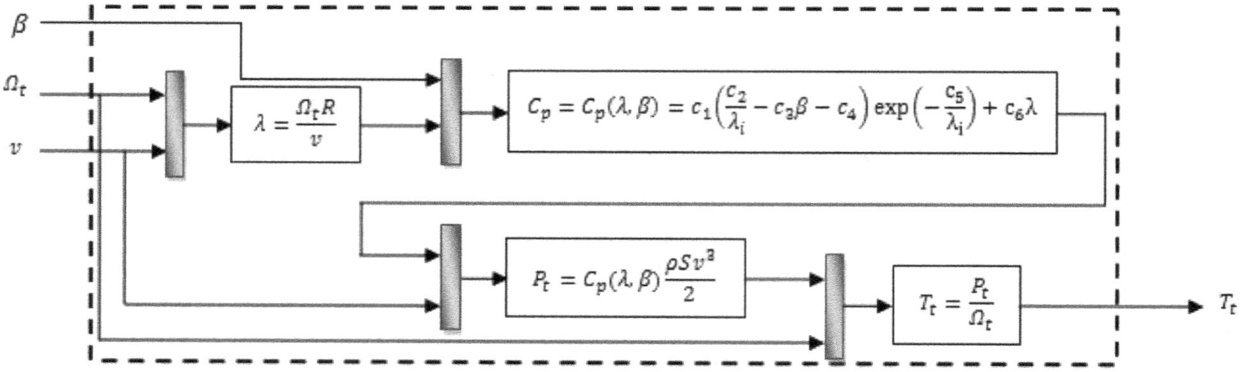

Figure 5. Modeling of wind turbine.

A relationship between C_p and γ for various values of the pitch angle β is illustrated in **Figure 4**. The maximum value of C_p, that is $C_{p-max} = 0.47$, is achieved for $\beta = 0$ and for $\lambda_{opt} = 8.1$. The optimal tip speed ratio $\lambda_{opt} = 8.1$ is a constant for a given blade. The equation 4 indicates that in order to obtain the maximum power and conversion efficiency, the turbine speed must be made adjustable according to the wind speed. The model of wind turbine is illustrated in **Figure 5**.

3. Induction generator model

The conversion of mechanical energy to electric energy is performed by the turbine and the generator. Different generator types have been used in wind energy systems. These include the squirrel cage induction generator (SCIG), doubly fed induction generator (DFIG), and synchronous generator (SG) [1].

The SCIG is simple and rugged in construction. It is relatively inexpensive and requires minimum maintenance. The SCIGs are also employed in variable-speed wind energy systems. To date, the largest SCIG wind energy systems are around 4 MW in offshore wind farms.

When the stator winding is connected to a distribution network, a rotating magnetic field is generated in the air gap of this machine. The rotating field induces a three-phase current in the rotor bars, which interacts with the rotating field to produce the electromagnetic torque as shown in **Figure 6**.

The voltage equations for the stator and rotor of the generator in the arbitrary reference frame are given by

$$\frac{d}{dt}\begin{pmatrix}\phi_{sa}\\ \phi_{sb}\\ \phi_{sc}\end{pmatrix} = \begin{pmatrix}v_{sa}\\ v_{sb}\\ v_{sc}\end{pmatrix} - \begin{pmatrix}R_s & 0 & 0\\ 0 & R_s & 0\\ 0 & 0 & R_s\end{pmatrix}\cdot\begin{pmatrix}i_{sa}\\ i_{sb}\\ i_{sc}\end{pmatrix} \quad (9)$$

and

$$\frac{d}{dt}\begin{pmatrix}\phi_{ra}\\ \phi_{rb}\\ \phi_{rc}\end{pmatrix} = -\begin{pmatrix}R_r & 0 & 0\\ 0 & R_r & 0\\ 0 & 0 & R_r\end{pmatrix}\cdot\begin{pmatrix}i_{ra}\\ i_{rb}\\ i_{rc}\end{pmatrix} \quad (10)$$

where

- ϕ_{sa}Z, ϕ_{sb}, ϕ_{sc}; ϕ_{ra}, ϕ_{rb}, ϕ_{rc}: stator and rotor flux-linkage vectors (Wb),
- v_{sa}Z, v_{sb}, v_{sc}: stator and rotor voltage vectors (V),
- i_{sa}Z, i_{sb},i_{sc}; i_{ra}, i_{rb}, i_{rc}: stator and rotor current vectors (A),
- R_sZ, R_r stator and rotor winding resistances (Ω).

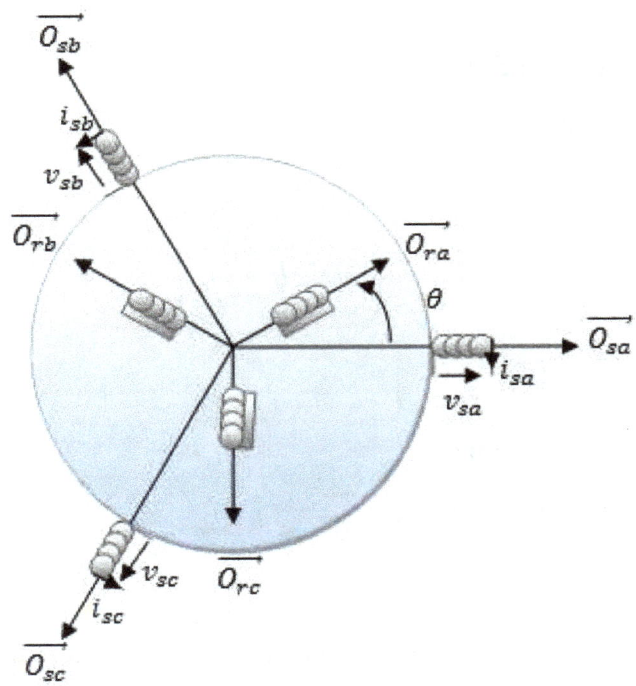

Figure 6. Representation of IG in electric space.

$$\frac{d}{dt}\left[\phi_{\mathrm{s}abc}\right] = \left[v_{\mathrm{s}abc}\right] - \left[R_{\mathrm{s}}\right]\cdot\left[i_{\mathrm{s}abc}\right] \quad (11)$$

and

$$\frac{d}{dt}\left[\phi_{\mathrm{r}abc}\right] = -\left[R_{\mathrm{r}}\right]\cdot\left[i_{\mathrm{r}abc}\right] \quad (12)$$

where

$$[I] = \begin{pmatrix} 1 & 0 & 0 \\ 0 & 1 & 0 \\ 0 & 0 & 1 \end{pmatrix} \tag{13}$$

The second set of equations is for the stator and rotor flux linkages:

$$\begin{pmatrix} \phi_{\mathbf{s}abc} \\ \phi_{\mathbf{r}abc} \end{pmatrix} = \begin{pmatrix} [L_s] & [M_{sr}] \\ [M_{rs}] & [L_r] \end{pmatrix} \cdot \begin{pmatrix} i_{\mathbf{s}abc} \\ i_{\mathbf{r}abc} \end{pmatrix} \tag{14}$$

where

$$[L_s] = \begin{pmatrix} l_s & m_s & m_s \\ m_s & l_s & m_s \\ m_s & m_s & l_s \end{pmatrix} \tag{15}$$

where

l_s: self-inductances for stator windings,

m_s: mutual inductances for stator windings.

and

$$[L_r] = \begin{pmatrix} l_r & m_r & m_r \\ m_r & l_r & m_r \\ m_r & m_r & l_r \end{pmatrix} \tag{16}$$

where

l_r: self-inductances for rotor windings,

m_r: mutual inductances for rotor windings

and :

$$[M_{sr}] = M_{max} \begin{pmatrix} \cos(p\theta) & \cos\left(p\theta - \frac{2\pi}{3}\right) & \cos\left(p\theta - \frac{4\pi}{3}\right) \\ \cos\left(p\theta - \frac{4\pi}{3}\right) & \cos(p\theta) & \cos\left(p\theta - \frac{2\pi}{3}\right) \\ \cos\left(p\theta - \frac{2\pi}{3}\right) & \cos\left(p\theta - \frac{4\pi}{3}\right) & \cos(p\theta) \end{pmatrix} \tag{17}$$

The third and final equation is the motion equation, which describes the dynamic behavior of the rotor mechanical speed in terms of mechanical and electromagnetic torque:

$$J\frac{d\omega_m}{dt} = T_e - T_m \tag{18}$$

The third and final equation is the motion equation, which describes the dynamic behavior of the rotor mechanical speed in terms of mechanical and electromagnetic torque.

3.1. Reference frame transformation

The transformation of variables between the three-phase stationary frame (abc frame) and the synchronous frame (dq rotating frame) is presented below.

3.1.1. *abc/αβ Reference frame transformation*

The transformation of three-phase variables in the stationary reference frame into the two-phase variables also in the stationary frame is often referred to as abc/$\alpha\beta$ transformation. Since the $\alpha\beta$ reference frame does not rotate in space

$$\left[x_{o\alpha\beta}\right] = [T_3] \cdot [x_{abc}] \tag{19}$$

where

$$T_3 = \sqrt{\frac{2}{3}} \begin{pmatrix} \frac{1}{\sqrt{2}} & 1 & 0 \\ \frac{1}{\sqrt{2}} & \frac{-1}{2} & \frac{\sqrt{3}}{2} \\ \frac{1}{\sqrt{2}} & \frac{-1}{2} & \frac{\sqrt{-3}}{2} \end{pmatrix} \tag{20}$$

3.1.2. *abc/dq Reference frame transformation*

To transform variables in the abc stationary frame to the dq rotating frame, simple trigonometric functions can be derived from the orthogonal projection of the *a*, *b*, and *c* variables to the *dq*-axis as shown in **Figure 7**.

The transformation of the *abc* variables to the *dq* frames, referred to as abc/dq transformation, can be expressed in a matrix form:

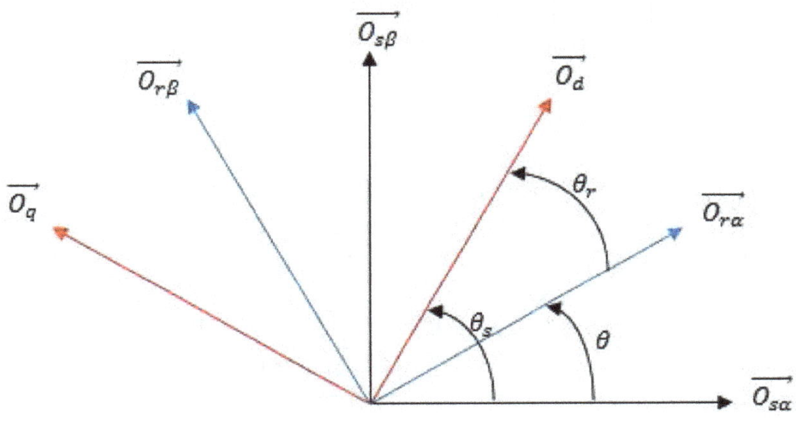

Figure 7. Transformation of (abc) stationary frame to the two phase (dq) arbitrary frame.

$$[x_{dqo}] = [P(\psi)] \cdot [x_{abc}] \tag{21}$$

where

$$[P(\psi)] = \sqrt{\frac{2}{3}} \begin{pmatrix} \cos(p\psi) & \cos\left(p\psi - \frac{2\pi}{3}\right) & \cos\left(p\psi - \frac{4\pi}{3}\right) \\ \sin(p\psi) & -\sin\left(p\psi - \frac{2\pi}{3}\right) & -\sin\left(p\psi - \frac{4\pi}{3}\right) \\ \frac{1}{\sqrt{2}} & \frac{1}{\sqrt{2}} & \frac{1}{\sqrt{2}} \end{pmatrix} \tag{22}$$

where

- $\psi = \theta_s Z$, for stator variables;
- $\psi = \theta_r Z$, for rotor variables.

So Eqs. (12) and (14) can be rewritten as

$$\frac{d}{dt}\left[\phi_{\mathbf{s}dqo}\right] = [v_{\mathbf{s}dqo}] - [R_{\mathbf{s}}] \cdot [i_{\mathbf{s}dqo}] - [\Theta] \cdot \left[\phi_{\mathbf{s}dqo}\right]\frac{d\theta_{\mathbf{s}}}{dt} \tag{23}$$

$$\frac{d}{dt}\left[\phi_{\mathbf{r}dqo}\right] = -[R_{\mathbf{r}}] \cdot [i_{\mathbf{r}dqo}] - [\Theta] \cdot \left[\phi_{\mathbf{r}dqo}\right]\frac{d\theta_{\mathbf{r}}}{dt} \tag{24}$$

where

$$[\Theta] = \begin{pmatrix} 0 & -1 & 0 \\ 1 & 0 & 0 \\ 0 & 0 & 0 \end{pmatrix} \tag{25}$$

the dq rotating frame, the flux, and currents are related by

$$\begin{pmatrix} \phi_{sdqo} \\ \phi_{rdqo} \end{pmatrix} = \begin{pmatrix} [L_{sP}] & [M_{srP}] \\ [M_{rsP}] & [L_{rP}] \end{pmatrix} \cdot \begin{pmatrix} i_{\mathbf{s}dqo} \\ i_{\mathbf{r}dqo} \end{pmatrix} \tag{26}$$

where

$$[L_{sP}] = \begin{pmatrix} l_s - m_s & 0 & 0 \\ 0 & l_s - m_s & 0 \\ 0 & 0 & l_s - m_s \end{pmatrix} \tag{27}$$

$$[L_{rP}] = \begin{pmatrix} l_r - m_r & 0 & 0 \\ 0 & l_r - m_r & 0 \\ 0 & 0 & l_r - m_r \end{pmatrix} \tag{28}$$

3.2. Back-to-back PWM converter

Power converters are widely used in wind power systems. (In fixed-speed, the converters are used to reduce inrush current during the system start-up, whereas in variable-speed, they are employed to control the speed/torque of the generator and also the active/reactive power to the grid) [1], where a back-to-back converter configuration with two identical PWM converters is used. The converters can be either voltage source converters (VSCs) or current source converters (CSCs)

In wind energy conversion systems, the converter is often connected to an electric grid and delivers the power generated from the generator to the grid. The converter in this application is referred to as a grid-connected or grid tied converter.

4. Control of wind turbine and induction generator

In variable-speed squirrel cage induction generator (SCIG) wind energy conversion systems, full-capacity power converters are required to adjust the speed of the generator in order to harvest the maximum possible power available from the wind. The generator-side converter (rectifier) is used to control the speed or torque of the generator with a maximum power point tracking (MPPT) (see **Figures 8** and **9**). The grid-side converter (inverter) is employed for the control of DC link voltage and grid-side reactive power. The system dynamic and steady-state performance are analyzed. The analysis is assisted by computer simulations and steady-state equivalent circuits.

4.1. MPPT control

The main goal of control is to maximize the wind power capture at different wind speeds, which can be achieved by adjusting the turbine speed in such a way that the optimal tip speed ratio $\lambda_{opt}t$ is maintained.

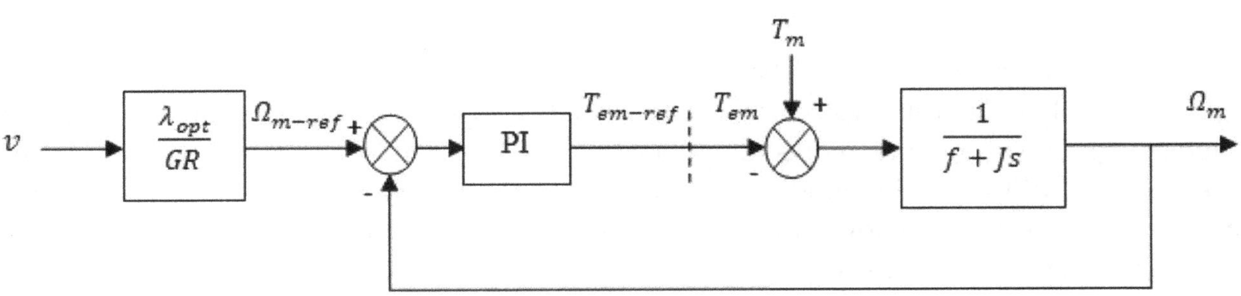

Figure 8. Control of mechanical speed Ω_m.

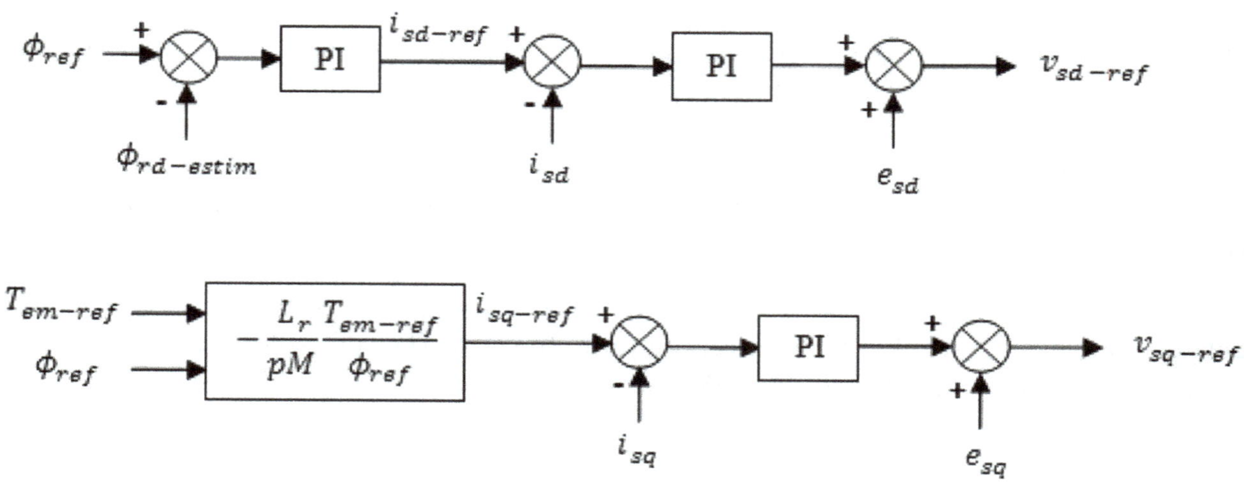

Figure 9. Vector control of SCIG.

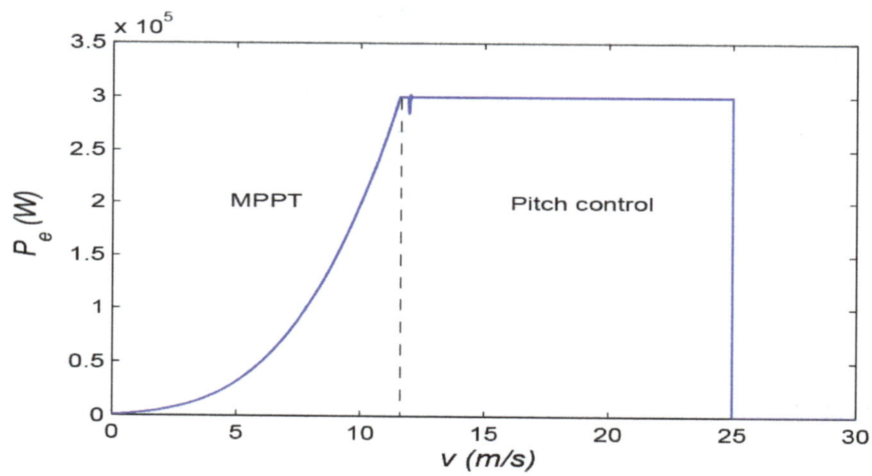

Figure 10. Wind turbine characteristic.

Control laws developed in this chapter are validated by simulations of the global wind behavior, applying a profile variable wind that can test these laws under and above the rated wind speed. This command allows to operate the wind so as to extract the maximum power of energy (MPPT) when the wind speed is below its nominal value and the limit above this speed (pitch control) as shown in **Figure 10**, and finally injecting a sinusoidal current in phase with the line voltage (PFC).

We note that the increase of β allows to degrade C_p coefficient and therefore causes the decrease in the mechanical power recovered on the axis of the wind turbine, simulation results are shown in **Figures 11**, **12** and **13**.

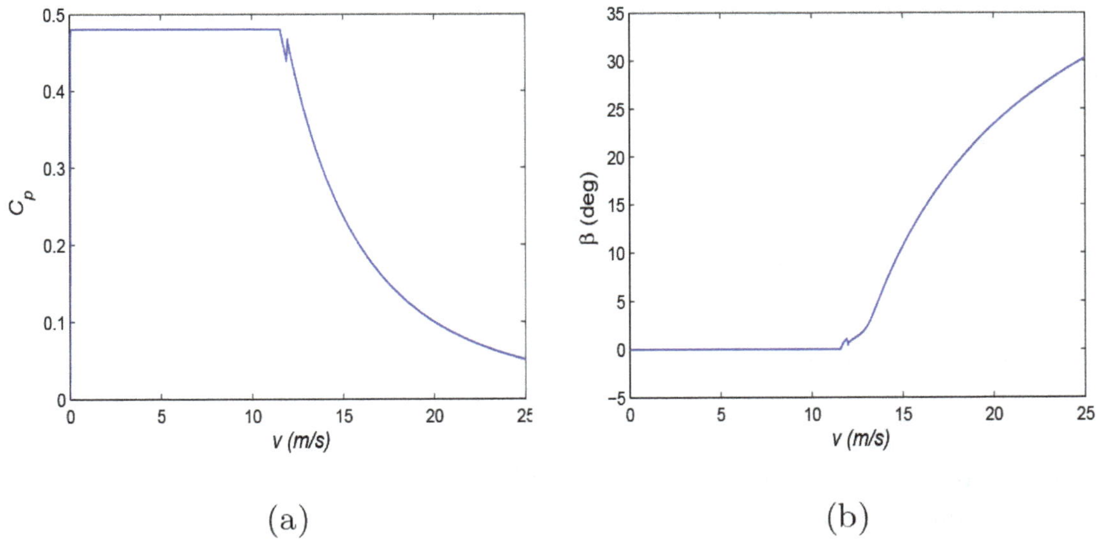

Figure 11. Variation of aerodynamic variables: (a) power coefficient C_p; (b) pitch angle β.

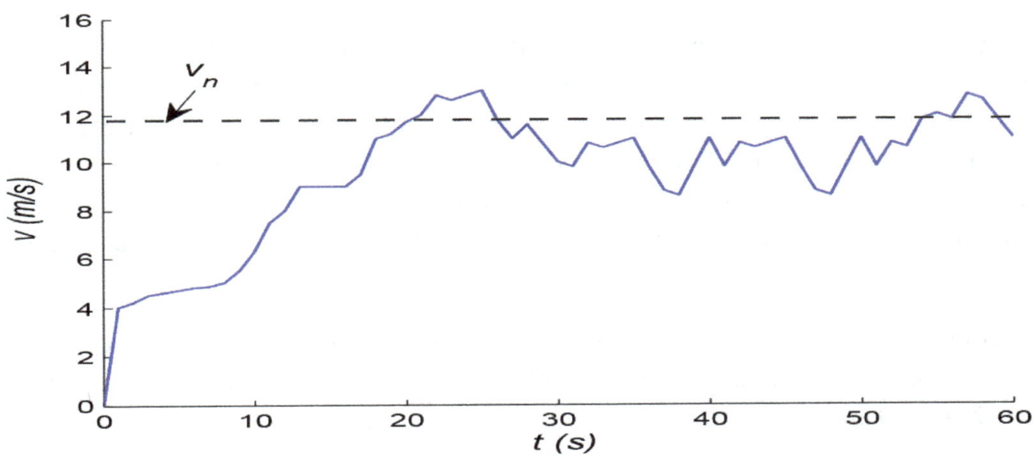

Figure 12. Wind velocity profile.

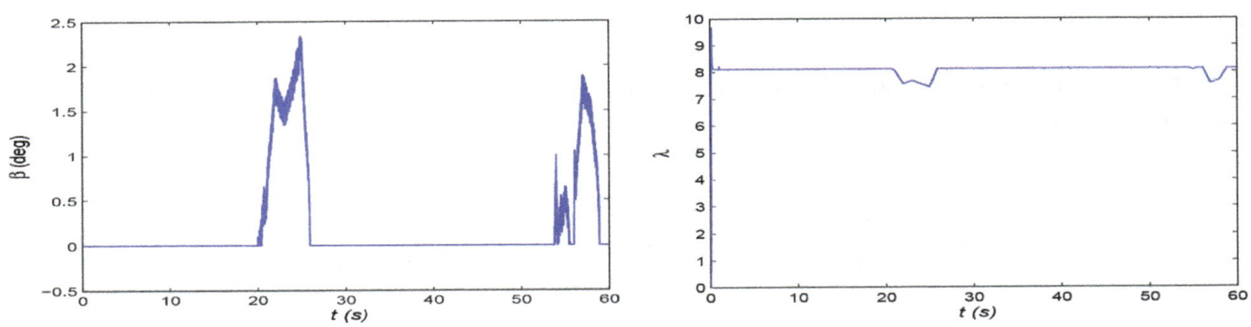

Figure 13. Impact on aerodynamic variables: (a) Pitch angle; (b) speed ratio β; speed ratio λ.

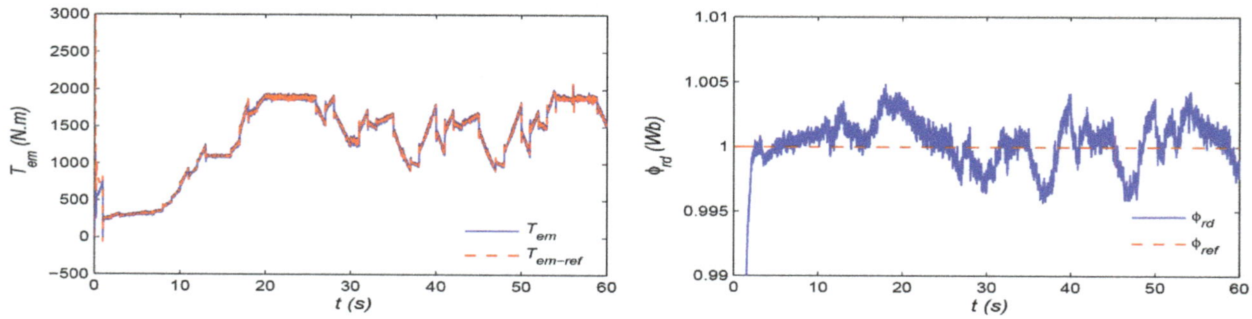

Figure 14. Variation of electromagnetic variables: (a) Electromagnetic torque; (b) Rotor flux.

Figure 14 presents the variation of electromagnetic torque and the rotor flux, as desired, they are tracked to their input references.

5. Control of grid-connected converter

A typical grid-connected inverter for wind energy applications is shown in **Figure 2**. The inverter is connected to the grid through a line inductance L_f, which represents the leakage inductance of the transformer. The power flow between the converter and the grid is bidirectional. Power can be transferred from the DC circuit of the converter to the grid. The grid power factor can be unity. It is often required by the grid operator that a wind energy system provide a controllable reactive power to the grid to support the grid voltage in addition to the active power production.

The grid-connected converter can be controlled with voltage-oriented control, as shown in **Figure 2**. According to **Figure 2**, the relationship between the grid, converter voltages, and line currents is given as follows:

$$\begin{pmatrix} v_{i1} \\ v_{i2} \\ v_{i3} \end{pmatrix} = R_f \begin{pmatrix} i_{g1} \\ i_{g2} \\ i_{g3} \end{pmatrix} + L_f \frac{d}{dt} \begin{pmatrix} i_{g1} \\ i_{g2} \\ i_{g3} \end{pmatrix} + \begin{pmatrix} v_{g1} \\ v_{g2} \\ v_{g3} \end{pmatrix} \tag{29}$$

Transforming the voltage Eq. (29) using dq transformation in the rotating reference frame at the grid frequency gives

$$\begin{pmatrix} v_{id} \\ v_{iq} \end{pmatrix} = R_f \begin{pmatrix} i_{gd} \\ i_{gq} \end{pmatrix} + L_f \frac{d}{dt} \begin{pmatrix} i_{gd} \\ i_{gq} \end{pmatrix} \begin{pmatrix} i_{gd} \\ i_{gq} \end{pmatrix} + L_f \omega_g \frac{d}{dt} \begin{pmatrix} -i_{gq} \\ i_{gd} \end{pmatrix} + \begin{pmatrix} v_{gd} \\ v_{gq} \end{pmatrix} \tag{30}$$

where ω_g is the speed of the synchronous reference frame. This indicates that the system control is cross-coupled, which may lead to difficulties in controller design and unsatisfactory dynamic performance. To solve the problem, a decoupled controller can be implemented. We consider the following couplings voltages:

$$\begin{pmatrix} e_d \\ e_q \end{pmatrix} = L_f \omega_g \frac{d}{dt} \begin{pmatrix} -i_{gq} \\ i_{gd} \end{pmatrix} \quad (31)$$

$$\begin{pmatrix} \Delta v_{fd} \\ \Delta v_{fq} \end{pmatrix} = R_f \begin{pmatrix} i_{gd} \\ i_{gq} \end{pmatrix} + L_f \frac{d}{dt} \begin{pmatrix} i_{gd} \\ i_{gq} \end{pmatrix} \quad (32)$$

The decoupled control makes the design of the PI controllers more convenient, and the system is more easily stabilized.

The equation for the voltage across the DC-bus is given by

$$C\frac{dU_{dc}}{dt} = i_s - i_g \quad (33)$$

This yields to

$$CU_{dc}\frac{dU_{dc}}{dt} = U_{dc}i_s - U_{dc}i_g \quad (34)$$

Eq. (34) can be rewritten as

$$CU_{dc}\frac{dU_{dc}}{dt} = P_s - P_g \quad (35)$$

From Eq. (35), the DC-bus voltage controller provides the active power reference value (or direct grid current reference) and Ps acts as a disturbance. The equation 34 can be presented as differential equations for DC-bus voltage and the grid currents [11].

To realize this control, the grid voltage is measured and its angle is detected for the voltage orientation. This angle is used for the transformation of variables from the abc stationary frame to the dq synchronous frame through the abc/dq transformation.

$$P_g = \frac{3}{2}\left(v_{dg}i_{dg} - v_{qg}i_{qg}\right) \quad (36)$$

$$Q_g = \frac{3}{2}\left(v_{qg}i_{dg} - v_{dg}i_{qg}\right) \quad (37)$$

To achieve the control scheme, the d-axis of the synchronous frame is aligned with the grid voltage vector; therefore, the d-axis grid voltage is equal to its magnitude ($v_{qd} = v_g$) and the resultant q-axis voltage v_{gq} is then equal to zero ($v_{gq} = 0$), from which the active and reactive power of the system can be calculated.

The initial angle of the phase 1 is set to 0 and, the initial angle of the d–q reference frame is set to $\frac{\pi}{2}$, and this causes the v_{gq} component to be zero and the v_{gd} component to be equal to simple phase voltage v_{g1}.

In this reference frame, the above P and Q equations will become [11]

$$P_g = \frac{3}{2} v_{dg} i_{dg} \tag{38}$$

$$Q_g = -\frac{3}{2} v_{dg} i_{qg} \tag{39}$$

Thus the active and reactive power flows are controlled by the i_{gd} and i_{gd}, respectively

5.1. Dynamic and steady-state analysis

The application of wind turbine power quality characteristics to determine the impact of wind turbines on voltage quality is explained through a case study considering a 3 MW wind farm on a 22 kV distribution network. To study the operation and performance of the SCIG wind energy system, some case studies are provided. One investigates the power dynamic behavior of the system, the impact on voltage and frequency, and another analyses the power quality performance of the system with a change in wind speeds. The control architecture of this conversion chain consists of different control blocks (see **Figure 15**).

5.2. Active and reactive power flow

The reactive power exchange with the grid is determined by the voltage and current of the grid side of the power electronic converter. In this case, the wind turbine is fully decoupled from the

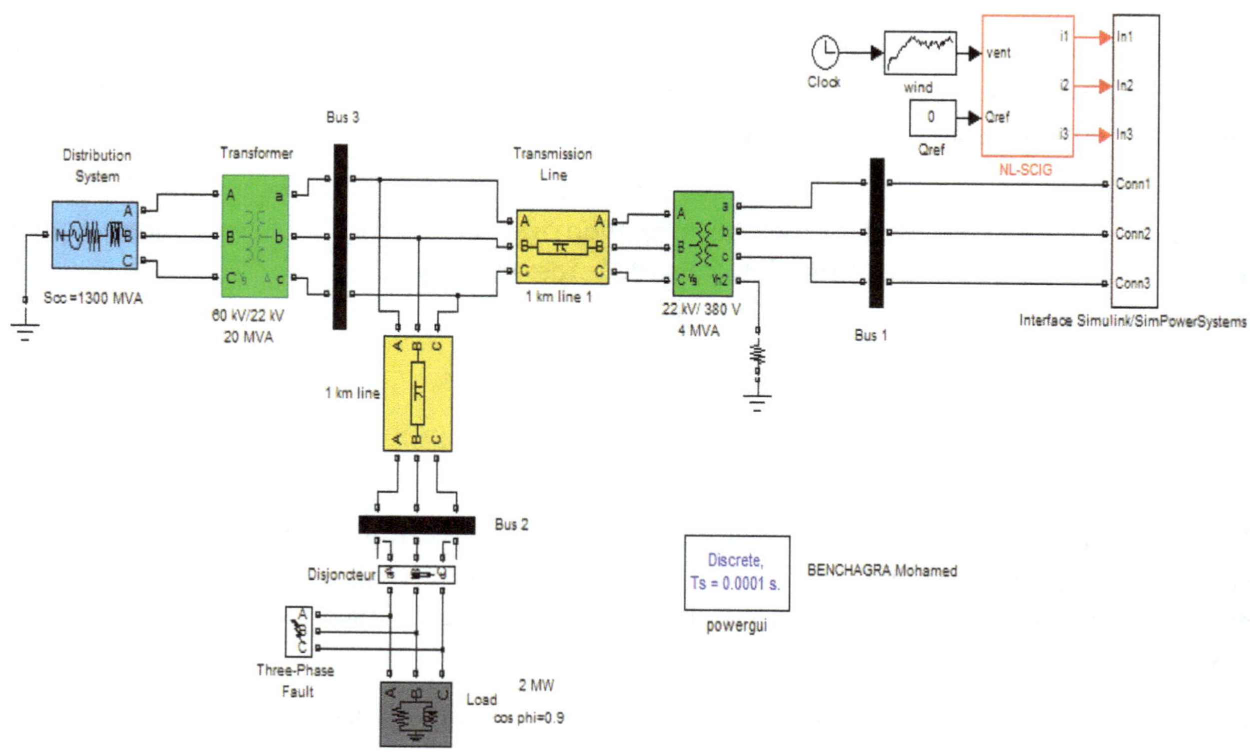

Figure 15. Simulation model of wind farm–connected distribution network.

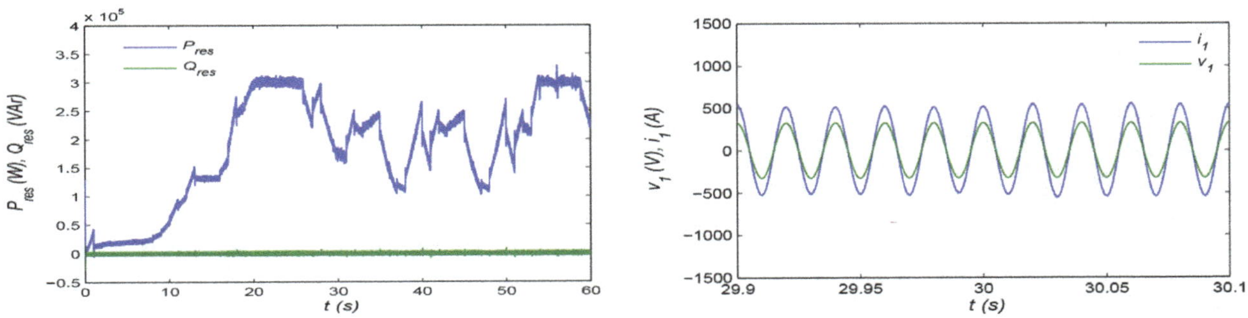

Figure 16. Variation of electric variables: (a) Active and reactive power; (b) voltage and current of wind turbine.

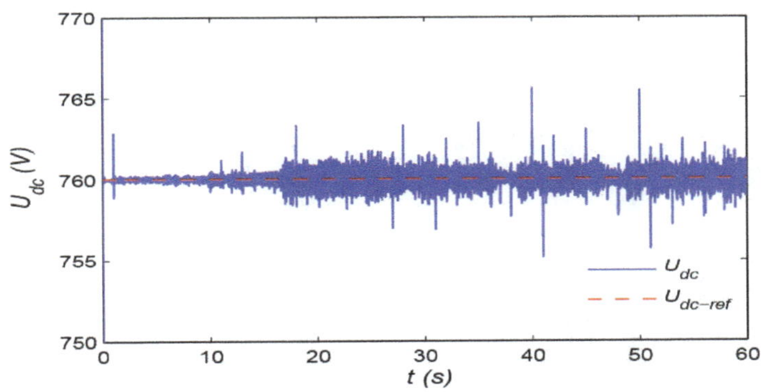

Figure 17. DC bus voltage.

grid. This means that the power factor of the wind turbine and the power factor of the power network side can be controlled independently. **Figure 16** shows the response of the system for a variable wind velocity. During test, the active and reactive powers are decoupled. **Figure 17** shows the DC-bus voltage of the back-to-back converter, as desired, the voltage is kept at their reference value.

5.3. Impact on voltage and frequency

The principal function of power network system is to transport and distribute electrical power from the generators to the loads. In order to function normally, it is essential that the voltage is kept close to its nominal value, so it is becoming increasingly important, for wind farm connected to the transmission system and for those connected to the distribution system, to be able to contribute to voltage control [12, 13]. This section looks at the impact of wind power on voltage control. However, even though there is a voltage difference between the two ends of the branch, the node voltage is not allowed to deviate from the nominal value of the voltage in excess of a certain value (normally 5% to 10%).

Ideally, the voltage should form a perfect sinusoidal curve with constant frequency and amplitude. However, in any real-life power system, grid-connected appliances will cause the voltage to deviate from the ideal [14] (**Figure 18 (a)**).

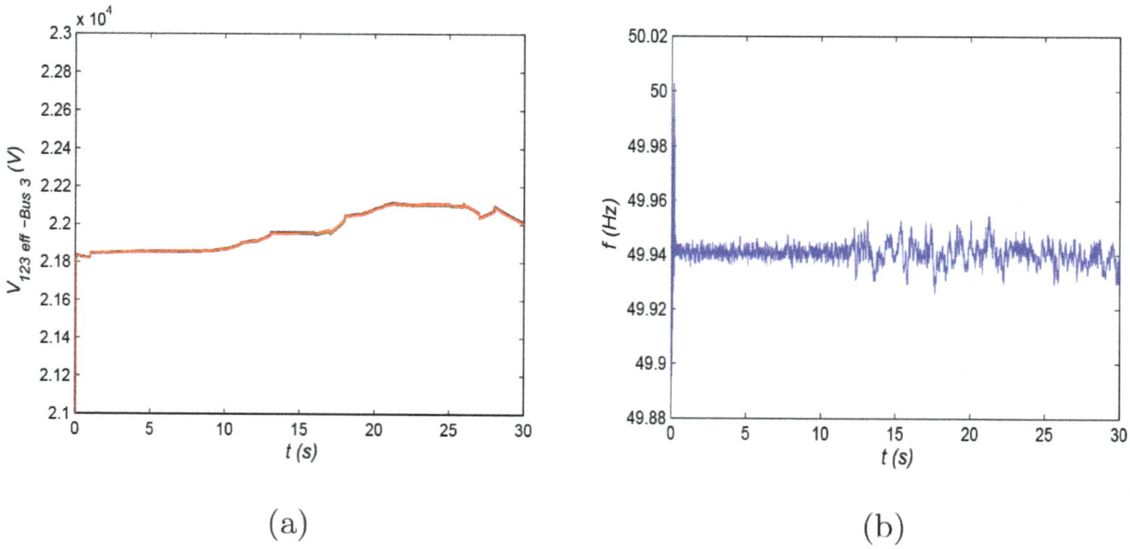

Figure 18. Impact on Bus 3: (a) voltage; (b) frequency.

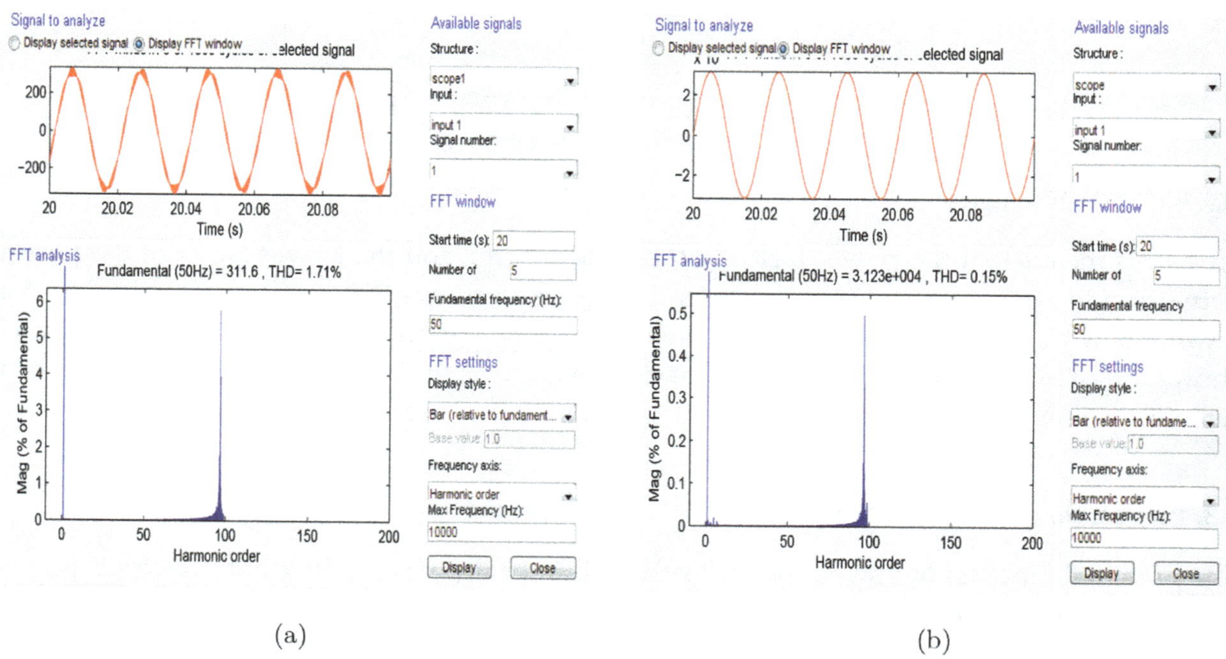

Figure 19. Harmonic analysis: (a) On Busbar 1; (b) On Busbar 3.

As can be seen, the adopted control strategy is capable of providing accurate tracking of voltage and frequency at normally values.

5.4. Power quality – harmonics

The application of power quality characteristics defined in IEC 61400–21 to determine the impact of wind turbines on voltage quality is explained in our case study by considering a

3 MW wind farm on a 22 kV distribution network. Certainly, if the wind farm is large or the grid is very weak, additional analyses may be required to assess the impact on power system stability and operation.

Nonlinear loads distort the voltage waveform and may in severe cases cause an overheating of neutral conductors and electrical distribution transformers.

The emission of harmonic currents and voltages of a wind farm with a back-to-back converter has to be stated. **Figure 19** illustrates the harmonic analysis on Busbar 1 and Busbar 3 of the proposed model. Its spectral analysis brings up by presence of the switching frequency (5 kHz) and a distortion (THD = 0.15%) of the effective value of the fundamental component of the voltage in Busbar 3 (Point of Common coupling). The results clearly show that the spread of harmonic current generated by wind turbines fit into the user standards.

6. Conclusion

This chapter discussed the impact of wind farm on the dynamics of distribution network. After modeling the system and developing the controllers, the obtained results are interesting for wind farm application to ensure better quality of power and to study the effect on power system dynamics or stability while increasing the wind power penetration.

Author details

Benchagra Mohamed

Address all correspondence to: mohamed.benchagra@uhp.ac.ma

University Hassan, ENSA, Khouribga, Morocco

References

[1] B. Wu, Power Conversion and Control of Wind Energy Systems, August 2011, Wiley-IEEE Press, ISBN: 978-0-470-59365-3.

[2] C. Chompoo Inwai, W.J. Lee, P. Fuangfoo, M. Williams, J.R. Liao, System impact study for the interconnection of wind generation and utility system, IEEE Transaction on Industry Applications, 41, 163–168, 2005.

[3] E. Hau, Wind Turbines: Fundamentals, Technologies, Application, Economics, 2nd ed., Springer Berlin Heidelberg, 2005, ISBN 978-3-540-29284.

[4] M. Benchagra, M. Maaroufi, M. Ouassaid, A performance comparison of linear and nonlinear control of a SCIG-wind farm connecting to a distribution network control, Turkish Journal of Electrical Engineering and Computer Sciences 22, 1–11, 2014.

[5] T. Burton, Wind Energy Handbook, John Wiley & Sons Ltd, 2011. ISBN: 978-0-470-69975-1.

[6] F. Blaabjerg, Z. Chen, Power Electronics for Modern Wind Turbines, Morgan & Claypool Publishers, 2006 doi:10.2200/S00014ED1V01Y200602PEL001.

[7] P.M. Pardalos, Handbook of Wind Power Systems, Springer, 2013. ISBN 978-3-642-41080-2.

[8] A. Baggini, Handbook of Power Quality, John Wiley & Sons Ltd, 2008 ISBN: 978-0-470-06561-7.

[9] M. Benchagra, M. Maaroufi, Control of wind farm connected distribution network, 10th IEEE International Conference on Networking Sensing and Control (ICNSC), 2013 10 -12 Apr 2013 , Evry, France.

[10] R. Penta, J.C. Clare, G.M. Asher, Doubly fed induction generator using back-to-back PWM converters and its application to variable speed wind-energy generation, IEE Proceedings – Electric Power Applications, 231–241, 143(3), 1996.

[11] M. Benchagra, Y. Errami, M. Hilal, M. Maaroufi, M. Cherkaoui, M. Ouassaid, New control strategy for inductin generator-wind turbine connected grid, 2012 International Conference on Multimedia Computing and Systems, 2012;10–12 May 2012, Tangiers, Morocco.

[12] D. Devaraj, R. Jeevajyothi, Impact of fixed and variable speed wind turbine systems on power system voltage stability enhancement. IET Conference on Renewable Power Generation (RPG 2011), 6–8 September 2011 - Edinburgh, UK.

[13] J.G. Slootweg, Wind Power and Voltage Control, Wind Power in Power Systems, 2005 John Wiley & Sons, Ltd , DOI: 10.1002/0470012684.

[14] J.O. Tande, Power Quality Standards for Wind Turbines, Wind Power in Power Systems, T. Ackermann editor, Wind Power in Power Systems, 2012 Second Edition, John Wiley & Sons, Lt, DOI: 10.1002/9781119941842.ch8.

4

The Importance of Internet of Things Security for Smart Cities

Mircea Georgescu and Daniela Popescul

Abstract

The purpose of this chapter is to provide an extensive overview of security-related problems in the context of smart cities. The impressive heterogeneity, ubiquity, miniaturization, autonomous and unpredictable behaviour of objects interconnected in Internet of Things, the real data deluges generated by them and, on the other side, the new hacking methods based on sensors and short-range communication technologies transform smart cities in complex environments in which the already-existing security analyses are not useful anymore. Specific security vulnerabilities, threats and solutions are approached from different areas of the smart cities' infrastructure. As urban management should pay close attention to security and privacy protection, network protocols, identity management, standardization, trusted architecture, etc., this chapter will serve them as a start point for better decisions in security design and management.

Keywords: Internet of Things, smart cities, Internet of Things security, attacks in Internet of Things, smart cities security

1. Introduction

During the history of mankind, cities have been trying to offer their residents a better quality of life, a safe and comfortable environment and economic prosperity. Nowadays, citizens expect from their cities fluid transportation, clean air, responsible consumption of utilities, constant interaction with city administrators, transparent governance, good health and educational systems and significant cultural facilities. In order to answer these requests, a city needs to become smarter and smarter, continuously improving its status quo. For the purpose of this chapter, we define a smart city as a future, better state of an existing city, where the use and exploitation of both tangible (e.g. transport infrastructures, energy distribution networks and natural resources) and intangible assets (e.g. human capital, intellectual capital of companies and organizational capital in public administration bodies) are optimized [1]. Summarizing the opinions expressed in [2–10], the relevant goals for a smart city are:

- Smart mobility (traffic management, bike/car/van sharing, multimodal transport, road conditioning monitoring, parking system, route planning, electric car gearing services);
- Smart grid/energy (power generation/distribution/storage, energy management, smart metering, street lightening optimization);
- Public safety (video/radar/satellite surveillance, environmental and territorial monitoring, children protection—e.g. safer home-school journeys for children, emergency solutions, waste management, smart air quality, weather data for snow cleaning);
- Smart governance (transparent decisional process, a greater involvement of citizens in legislative initiatives, public-private partnerships, online taxing systems);
- Smart economy (high-level jobs, competitiveness, entrepreneurial spirit, innovation and research in the field) and
- Smart life (cultural and educational facilities, meaningful events, entertainment and guided tours, access to cultural sights and historical monuments, good conditions for health).

An essential element of a smart city, often neglected when focus is placed on infrastructure, is the self-decisive, independent and aware citizen. In [11], humans are seen as sensors, with a direct and active public participation, strongly facilitated by information and communication technologies (ICT). According to [12], the relationship between the city and the smart citizen should be characterized by urban openness, defined as systems' capacity to enable user-driven innovation in existing and new services, participatory service design and open data platform availability. Also, service innovation, partnership formation and urban proactiveness (the extent to which smart city services are moving towards sustainable energy use as well as ICT-enabled services) are mandatory.

In recent years, the fulfilment of these goals depends more and more on technology, especially ICT. In consequence, one of the essential nuances of the term "smart city" is given by the ICT incorporation in urban infrastructure, with solutions as city operating systems, centralized control rooms, urban dashboards, intelligent transport systems, integrated travel ticketing, bike share schemes, real-time passenger information displays, logistics management systems, smart energy grids, controllable lighting, smart meters, sensor networks, building management systems, various smartphone apps and sharing economy platforms, etc. [12–15].

Internet of Things (IoT) has a central place among these technologies. In IoT, the physical things connect to other physical and virtual things, using wireless communication and offering contextual services. IoT is based on a global infrastructure network which connects uniquely identified objects, by exploiting the data captured by the sensors and actuators, and the equipment used for communication and localization. The radio-frequency identification

(RFID) lies at the basis of this development, but the IoT has developed by incorporating technologies such as sensors, printed electronic or codes, PLC, EnOcean, GPS, mobile (2G/GSM, 3G, 4G/LTE, GPRS) and short-range (NFC, Bluetooth, ZigBee, Wi-Fi, ANT, Z-Wave, IEEE 802.15.4) communications. The collaboration of the cyber-real artefacts is changing the city infrastructure, and their autonomous and nomad characteristics might lead to serious security problems that must be understood and solved in good time. A key challenge for IoT towards smart city applications is ensuring their reliability, incorporating the issues of ethics, security (confidentiality/integrity/availability), robustness and flexibility to rapidly changing environmental conditions. Without guarantees that the interconnected objects are accurately sensing the environment and are exchanging the data and information in a secure way, users are reluctant to adopt this new technology. The people's trustful acceptance of IoT components in a smart city is closely related to the notions of risk, security and ensuring private life which must be properly addressed by urban management.

2. Security challenges in Internet of Things

The aspects related to ethics and security in ICT have been a subject of study for the academic world and the wide public since the appearance of computers and the prefiguration of artificial intelligence. Thus, it is said that ICTs are of an emergent and creative nature and, explicitly or implicitly, they overtake some of our tasks and delicately induce certain moods or even force behaviour patterns, following their own development and functioning logic, imperatively heading the humankind to its maximum efficiency. Society can only answer to this by adapting and accepting the situation. Over the time, security in ICT has been treated from a historical perspective, at the organizational level, from a hacker's point of view or from a technical one. Currently, researchers approach the so-called green technologies, calm technologies, cloud computing, the impact of social media on people and communities and especially IoT, which raises a great number of security questions.

Difficulties in approaching IoT security are brought at least by the following elements:

- While city security is addressed primarily by city managers, IoT is rather understood by engineers. These two sides must dialogue and transfer knowledge both ways, a process which is not necessarily easy. If the authors of norms, standards, programs and security policies lag behind technical experts, the digital divide may deepen a lot and collaboration may prove difficult.
- One of the information security truisms says that the attackers are always one step ahead the "good guys". But while current, "classical" Internet attacks may cause damages to the information confidentiality, integrity and accessibility, similar actions in IoT can lead even to the loss of human lives. As shown in [16], there have already been demonstrations of hackers' interferences in the on-board computers of cars/planes and attacks in surgery rooms or on patients with implanted insulin pumps or other medical devices. As the list of vulnerable systems includes electric heating systems, food distribution networks, hospitals, traffic lights systems, transport networks, which are strongly interconnected in a smart city,

the attack scenarios which might be envisaged starting from here are truly scaring. In consequence, the importance of security measures increases greatly in the IoT.

- Besides attackers, the autonomous behaviour of things that invisible communicate to each other can affect our lives, in ways still difficult to predict. Anticipating dangers in IoT through a serious vulnerability scan becomes a necessity, but the process is difficult and can be done only with a sustained research and practice effort.
- IoT landscape is fragmented, because its applications are based on different architectures, standards and software platforms of significant complexity. Each smart city develops proprietary technological solutions, in response to its own problems and opportunities. In many situations the connected things, technologies and their firmware are protected by trade secrets. Legal framework is not yet appropriate, and legal responsibilities are not clear enough. Existing solutions are not interconnected and standardized, creating so-called technological silos; also, a lot of actors are involved, and various regions of the systems are controlled by different organizations.

Even this non-exhaustive presentation of the IoT-related security issues is an alarm sign that, in a smart city, every inhabitant should be assured he/she is protected by efficient technical, economic, legal and social actions. In what follows, the above mentioned problems are going to be approached in a framework in which smart cities are seen as a synergetic sum of smart devices that generate huge amounts of data while working for the smart citizens' benefit.

2.1. Security vulnerabilities in Internet of Things

The most important vulnerabilities in IoT are determined by the special nature of interconnected objects and the great variety and sensitiveness of the data collected.

2.1.1. Not-so-smart things

The objects interconnected in IoT and used in smart cities are characterized by ubiquity, miniaturization, autonomy, unpredictable behaviour and difficult identification. Their heterogeneity is impressive, ranging from tiny/invisible objects to very sophisticated embedded systems. In the same city, we can easily identify sensors used to monitor pollution and air quality, traffic and the greater road infrastructure, public and private safety, energy and water consumption, waste management, etc.; wearable sensors, placed into clothing or under the skin; usual things such as keys, watches, coffee filters, fridges, domestic heating controllers, books, doors, etc. and devices with a lot of computing power such as smartphones, tablets, printers, TVs, medical devices, SCADA (supervisory control and data acquisition) systems, cars, etc. Their number increases on a daily basis, and so do the connections between them. According to [16], all these things can be very smart in some situations and quite stupid in others: for example, smart in the sense that they collect, transmit, process and respond to various data, but stupid when there is a need to protect them. In [17], software, hardware and network constraints that restrict the inclusion of adequate security mechanisms (e.g. cryptography) directly in smart objects are identified. For this reason, security measures are usually left aside, and the exposure to attacks is high. A Hewlett-Packard study is mentioned in [18] —it shows that 80% of things in IoT fail to require passwords of a sufficient complexity and length, 70% enable an attacker to identify valid user accounts through account enumeration, 70% use unencrypted network services and 60% raise security concerns with their user interfaces.

2.1.2. Deluges of sensitive data and information

Data collected by smart things are at the heart of smart cities. The problem is that they are sensitive data, often gathered without citizens' explicit consent. For example, messages, medical and academic records, personal pictures, appointments, bank account information, contacts and others can be used by the smart cities' infrastructure, with more or less security measures put in place. Safely combining IoT data from different sources is a serious issue in a smart city, since there is no guaranteed trusted relationship between the parties involved. As regards the property right on data and information, the difficulties appear from the correct identification of the authors—for example, an answer to the question "Who is the owner of data retrieved by sensors connected in IoT?" is hard to imagine at this point. When the information is personal or financial, things get more serious. The IoT omnipresence will make the boundaries between the public and private space invisible, and people will not know where their information security ends up. The Big Brother type surveillance, namely monitoring the individuals without them being aware of it, will be possible.

User privacy is strongly affected by the fact that the objects are equipped with sensors which will allow them to "see", "hear" or even "smell". The data registered by the sensors are sent in great quantities and in different ways through networks, and this can prejudice the individual's private life. According to [19], today's average smart mobile devices and applications are capable of recording user mileage, blood pressure, pulse and other intimate medical data that can be stored or sent to points of interest without the explicit user consent. These facts combined with the estimate that in 2020 the number of interconnected devices from IoT will exceed 25 billion can have devastating consequences. By means of RFID, GPS and NFC technologies, the geographic position of where a person is and his/her movements from one place to another can be easily found without his/her knowledge.

At a supra-level, smart spaces want to know everything about their inhabitants. As presented in [12], various technologies capture personally identifiable information and household level data about citizens (their characteristics, their location and movements and their activities), link these data together to produce new derived data, and use them to create profiles of people and places and to make decisions about them. For example, a smart building is sensitive in terms of environmental condition (temperature, humidity, smoke, CO_2, extreme light, air pollution, external presences) and is also able to determine a very accurate user profile based on his/her habits. Vehicles are active members of cities; they interact with each other, with drivers/passengers and with pedestrians. As shown in [19], they have embedded computers, GPS receivers, short-range wireless network interfaces and potentially access to in-car sensors and the Internet. The smart city infrastructure can read data about vehicles using radars, Bluetooth detectors and license plate cameras. Speed, flow and travel times are known this way and they can be associated with the driver's identity. According to [20], tracking can reveal sensitive locations, such as home or work locations, along with the time and duration of each visit, effectively allowing one to infer the detailed behavioural profiles of drivers, information about safety-critical events, speed, destination, home and workplace addresses, time spent in a particular location and so on.

2.2. Security threats in Internet-of-Things

Security threats can be divided, according to their nature, into three major categories: natural factors, based on hazard; threats caused by incidents that appeared in the system (errors); threats on systems caused by human-intended action (attacks).

2.2.1. Natural factors

The natural causes based on hazard, that can affect the IoT implementations in a smart city, can be divided into *special environment conditions* and *natural calamities or disasters*. The first category includes extremely high or low temperatures, excessive humidity or an excessively dusty environment which, in time, can determine IoT devices to break down. In the second case, the smart city infrastructure can be affected by fires, floods, strong winds, storms or earthquakes.

2.2.2. Incidents/errors

One of the most frequent *human errors* that can emerge when using IoT devices is the improper configuration, ignoring the activation of the login function or of other security mechanisms. The devices are not configured in an adequate manner, implicit factory settings are used and this is especially dangerous when passwords are involved. Proper authentication settings are not put in place, terms and conditions are not read/understood and there is no knowledge about the data collected by applications and the way of using them by third parties. Also, people give the same treatment to all the data stored in the device—without taking into account the fact that certain data, when loaded onto IoT devices, can require extra security measures. Unaware citizens are easily fooled through social engineering, spam emails, data streaming and other malicious methods. More severe are the errors that appear in the configuration of networks. The causes of errors are the "classic" ones—insufficient qualification/thoughtlessness, people's involvement in problems that are out of their competences (either due to curiosity, or from an exaggerated reliability in their own power to solve certain things), ignorance (we shouldn't expect users to use a system correctly if they haven't been trained to do so) and lack of interest in performing certain actions.

The *problems related to the software* are much more numerous in the IoT environment as compared to the classical environment, as a result of the juvenile character of IoT applications. Producers have difficulties in developing software which functions properly on all customized models. Even more challenging is the problem of portability for those who develop software for the whole range of devices found on the market. The significant software complexity involved by IoT, the requirement that each object/device must have a unique identity and the large code base cause difficult testing and validation procedures. In a more specific manner, [21] shows that encryption is not used to fetch updates, update files are not properly encrypted, updates are not verified before upload and firmware usually contains sensitive information.

For various reasons, the services offered by IoT providers do not function in normal terms all the time and *communication line breakdowns/lack of signal/connexion errors* occurs. A malfunctioning at the level of a network, either from a provider or from within an organization, can result in the blocking of the infrastructure in a certain area of the city. Wireless networks are more vulnerable than the wired ones, due to interferences, frequent disconnections, broadcast transmission of data, low capacity and great mobility of devices. In consequence, the wireless channels are more susceptible to errors and this may lead to the degradation of security services, easier data interception and difficult use of advanced encrypting schemes. The physical security of objects is not guaranteed and their identification and authentication are problematic, especially in the public networks; the control of the objects may be lost and cascade failures may appear, caused by the interconnectivity of a large number of devices, difficult to be protected simultaneously.

2.2.3. Attacks

In a smart city, the attack surface is an extended one. Usual problems refer to device deliberate damage/theft, attacks on devices/components intended for recycling, malware and phishing attacks, network spoofing attacks or social engineering (e.g. apps repackaging—a malware writer takes a legitimate application, modifies it to include malicious code, then sets as available for download—or attacks using a newer version of software—creator of the malicious software sets a newer version of the app, infected with malware to the smart device user). But there are also numerous novel problems that make the attack scenarios inexhaustible.

First of all, we notice a large and increasing number of *sensor-based attacks*. To start from our pockets, we must admit that the inventory of sensors in a smartphone is intimidating: GPS chips, microphones, cameras, accelerometers, gyroscopes, the proximity sensors, magnetometers, ambient light sensors, fingerprint scanners, barometers, thermometers, pedometers, heart rate monitors, sensors capable to detect harmful radiation, back illuminated sensor, RGB light sensors, hall sensors [22]. Such sensors detect location of the mobile phone, in this way helping users to navigate in cities by maps/pictures, measure the position, tilt, shock, vibration and acceleration (the rate in change of velocity), rotations/twists, detect the presence of nearby objects without any physical contact, capture how bright the ambient light is, measure atmospheric pressure, deliver altitude data, detect the minute pulsations of the blood vessels into one's fingers and calculate one's pulse. They can capture location, movements, time stamps, even private conversations and background noises. As a result, a smartphone can be used to keep a targeted individual under surveillance. This, combined with the possibility of installing third-party software and the fact that a smartphone is closely associated with an individual, makes it a useful *spying tool*.

From a different point of view, the use of these sensors by different applications, the quantity and the purpose of collected data are not fully understood and controlled by their owners. For example, as shown in [23], video and pictures can reveal the social circle and behaviour of a citizen in a completely unexpected manner; in addition, according to [24], smartphones are more and more targeted by *malware* which accesses the microphone, cameras and other sensors. The book mentions Soundcomber, a proof-of-concept Trojan horse application that records the sounds made when digits are pressed, identifies them and tries to reveal typed PINs or passwords.

In another academic demonstration described in [25], when users placed their smartphone next to the keyboard, the deviations of accelerometer were measured. In this way, entire sequences of entered text on a smartphone touch screen keyboard were intercepted. In [26] and [27] similar successes are presented: using the motion sensors (accelerometers and gyroscopes), keystrokes (four-digit PINs and swiping patterns) were inferred from touch screens of smartphones and tablets with various operating systems. Also, in [28] it is showed that the gyroscope can be used to eavesdrop on speech in the vicinity of the phone.

From another range of IoT devices, thermostats communicate their location (including the postcode), temperature data, humidity and ambient light data, the time and duration of activation—these data can be used to determine domestic habits of a citizen; medical bracelets store the heartbeat and sleeping patterns, collecting biometric and medical data that reveal individuals' physiological state. It is obvious that if these valuable data are not well treated, significant privacy problems may occur.

Various new attacks are also permitted by *short-range communication technology*. ZigBee is a global standard and protocol developed as a light wireless communication for helping the smart objects to address one to each other in a common and easy way. With low costs and good efficiency, ZigBee technologies are used in many scopes such as home automation, industrial control or medical data collection. ZigBee-enabled systems are vulnerable to security threats, such as traffic sniffing (eavesdropping), packet decoding and data manipulation/injection. Moving on to Bluetooth, some blue-prefix attacks are bluejacking (spamming nearby object users with unsolicited messages), bluesnarfing (stealing the contact information found on vulnerable devices) and bluebugging (accessing smart objects' commands without notifying or alerting their user). Also, anyone with a Bluetooth-enabled device and software for discovering passwords via multiple variants (brute force) could connect to road sensor, etc. Regarding Near Field Communication, possible security attacks include eavesdropping, data corruption or modification, interception attacks and physical thefts. At a 2012 BlackHat conference, a researcher presented his findings on how he hacked smart devices to take advantage of a variety of exploits [29].

2.3. Living in a smart city—some risky scenarios

If we take into consideration the smart cities' dimension, we can imagine a multitude of scenarios as effects of the previously mentioned vulnerabilities and threats.

According to Bettina Tratz-Ryan, research vice president at Gartner, "smart commercial buildings will be the highest user of IoT until 2017, after which smart homes will take the lead with just over 1 billion connected things in 2018" [30]. *Smart buildings* increasingly use technology to control aspects such as heating, lighting and physical access control—all of which are potential vectors for attackers to target. A building automation system (BAS) controls sensors and thermostats. Several areas of concern were found in the BAS architecture that could allow hackers to take control, not only of the individual building system but also of the central server, which could then be a springboard to attack other buildings. After this proof-of-concept, IBM X-Force ethical hacking team leader Paul Ionescu said that the exercise proved that very little attention was being paid to IoT in smart buildings as these devices fell outside the scope of traditional ICTs [31].

In an attempt to explore security issues in *smart city transport infrastructure* and give recommendations on how to address them, presented in [32], a Kaspersky Lab Global Research & Analysis Team (GReAT) expert has conducted field research into the specific type of road sensors that gather information about city traffic flow. Team demonstrated that information gathered by these devices, delivered and analysed in real time by the special city authorities, can be intercepted and misused, in scenarios as demolishing expensive equipment and sabotaging the work of the city authority's services. In [33], some attacks which enable the hackers to stop the engine during the travel or opening the doors of the car into the parking lot are presented. From a different point of view, [34] showed that, in public transportation, screen reflected in sunglasses were filmed and, with a special software, password entered by users were discovered.

Another example in [34] demonstrates that the mobile infrastructure used by *the police forces in a smart city* is vulnerable. With low costs and large-available equipment (including a GirlTech IMME toy instant messenger of 15$), denial-of-service and interception attacks were proved as possible. Captured clear text data included identifying features of targets and undercover agents, plans for forthcoming operations, wide range of crimes, etc.

Denial-of-service attacks can be trivially launched by malicious entities against a wireless-based communication infrastructure. In the context of a *smart grid*, such attacks have potential to disrupt smart grid functions such as smart metering, demand response and outage management, thus impacting its overall resiliency [35].

In the *health area*, [24] presents a science-fiction scenario, in which Brain-computer interfaces (BCI)-based games could provide their users stimuli that generate subconscious thoughts (e.g. part of a PIN number, passwords, financial data). These thoughts are captured by the BCI device and sent to the attacker, who analyses them, searching for sensitive information.

As presented in [16], attacks in these zones can provoke compromising entire systems, and an infection can be easily transmitted between systems. This, in extremis, can determine an infection of the city itself, destroying even the physical infrastructure and threatening lives. This scenario seems to be a science-fiction one, but it's important to remember that Stuxnet, an "unprecedentedly masterful and malicious piece of code", has been sold on the black market since 2013. The experts in ICT security say it could be used to attack any physical target which is related to computers, and the list of vulnerable systems is almost endless—electric heating systems, food distribution networks, hospitals, traffic lights systems, transport networks, etc. Another malware, such as Linux.Darlloz worm, infects a wide range of home routers, set-top boxes, security cameras and other consumer devices that are increasingly equipped with an Internet connection. In these conditions, the terrorist cyber-strikes against the utility and industrial infrastructure can no longer be dismissed as a spy movie scenario. In an analysis on industrial control systems (SIEMENS S7, MODBUS, DNP3, BACNET) security made at Romania's level, [35] showed that most vulnerabilities were found in GSM towers, utilities providers, furnaces and data centres. Intrusions in SCADA systems can lead to disruptions in the exchange of data between control centres and end-users. As a result, certain services provided to citizens (access to public health services in critical moments, the supply of electricity in some areas) will be compromised; certain areas of the city can be blocked by stopping traffic lights, etc. Intruders can also install malware systems in data centres/user devices to obtain sensitive information about citizens and to use them for criminal purposes.

3. IoT-related security measures for a safer smart city

In an IoT-based smart city architecture, development and progress are not possible without trust. Security of each device, sensor and solution is not optional; it definitely must be taken into consideration from the very beginning. On the above presented quicksands, the need to rethink the "classical" security measures appears as mandatory. Also, specific novel measures are needed from various actors.

3.1. Legal/governmental actions

Through vast regulations and proper financing, European Union (EU) made an impressive start in the smart cities' security field. EU leaders affirm that security should play an important role in any smart city development strategy, taking into consideration those web-based attacks in IoT increased by 38% in 2015 [36]. Alliance for Internet of Things Innovation (AIOTI), an organization founded by the European Commission and various IoT key players in 2015, strongly recommends the principles of "privacy by design" (inclusion of proper security measures at the earliest stage in technological design) and "privacy by default" (no unnecessary data are collected and used) [37]. Under this umbrella, partners with different backgrounds—local authorities, telecom operators, universities, companies, small and medium enterprises—bring together their complementary legal, academic, societal, technical and business expertise and implement powerful projects. Some of the (intended) results of selected projects are presented in **Figure 1**.

Also, most European government affirm a strong interest in securing IoT, which is, in their opinion, an important factor for innovation and growth.

3.2. City managers

In a smart city, programs, policies, procedures, safety standards, best practices, security incidents and event management systems need to be developed and put in place. This is the attribution of the city administrator; cooperation with private sector is also mandatory. Proper audit trail mechanisms are needed in order to ensure that no limits are crossed by service providers. Because the smart cities grow, the infrastructure becomes more interconnected and risks are multiplying. A coherent and stable digital architecture must be maintained. By identifying vulnerable systems, assessing the type and magnitude of probable risks and instituting remedial measures, these bodies can fight cyber-physical-attacks and create risk-resilient smart services, maintaining the trust of their inhabitants that systems are safe and secure.

Figure 1. Smart city–related security results in EU-funded projects.

ICT departments of the public administration have to educate the citizens in a proper way. They can use social media tools in order to provide increased awareness and control and to empower citizens to easily manage access to IoT devices and information, while allowing IoT-enabled, citizen-centric services to be created through open community APIs. No doubts regarding the collection of data and misunderstandings of legal framework are allowed to occur—inhabitants must be informed directly of any risk related to their privacy and security. Secure exchange of in-transit and at-rest data is required between IoT devices, cities and citizens. The ultimate goal is a more self-aware behaviour of users, e.g. use of two steps of authentication on devices—at minimum, default passwords should be replaced with stronger ones; password encryption, or constant software updates.

3.3. Producers/security providers/software developers

Producers have to provide secure design and development of hardware—security methods should be built into the IoT equipment and network at the very beginning of the process, and not after its implementation. The cooperation with security providers/researchers is mandatory—they need to adapt the "classical" security methods as encryption, identity management techniques, device authentication mechanisms, digital certificates, digital signatures and watermarking to the new environment, and to make them available for all entities interested in a proper data protection, also they can help producers to find and patch all the vulnerabilities before it's too late.

At the device level, information about the default names, MAC and IP addresses, ports, technological processes used in production phase, even the producer/vendor's name should be kept confidential; if the attacker has this information, he can easily find online tools for hacking the device and can obtain control on management systems of smart infrastructure. Better user configuration capabilities are necessary, as the number and the complexity of systems make it necessary to provide mechanisms allowing the users to configure the systems themselves. Feedback should be required from the users in a coherent way; consumers' opinion must be taken into consideration when devices/networks are redesigned.

In software development, testing should receive proper attention—good security scanning before launching the code is a common sense request. Also, better controls on who has access to software are needed, preventing leakage of information about passwords. Application developers need to specify in a very clear way the measures they have taken before user's private and confidential data are accessed, and the anonymizing and encryption procedures used when data are in transit.

4. Conclusions

In a smart city, IoT interferes strongly with inhabitants' lives. IoT, which is no more in its infancy, presents various vulnerabilities and threats, caused by technological advances and proliferated through lack of users' awareness. They are augmented by the extended use of new technologies as RFID, NFC, ZigBee, sensors, 3G and 4G that bring along the adjustment of the traditional information security threats to this new environment, as well as the emergence of new dangers. The problems treated here are of interest both for each of us, as citizens, and for the city managers, national and international regulators, especially in a world in which the borderline between the physical and virtual life is becoming more and more difficult to draw.

In this context, urban managers have to address carefully the notions of trust, risk, security and privacy. The city authority have to be well informed about all the problems related to smart things, spaces, services and citizen security; also, the solutions offered by the security providers have to be known and chosen with maximum discernment.

The chapter offers only a non-exhaustive review of vulnerabilities, attacks and security measures, with the intention to raise awareness in this area of large public interest. Further in-depth analyses for each vulnerability, attack scenario and security measures adequacy are necessary.

Author details

Mircea Georgescu* and Daniela Popescul

*Address all correspondence to: mirceag@uaic.ro

"Alexandru Ioan Cuza" University, Iași, Romania

References

[1] Georgescu, M, Păvăloaia, VD, Popescul D, Țugui, A. The race for making up the list of emergent smart cities. An Eastern European country's approach. Transformations in Business & Economics. 2015;14(2A (35A)):529-549.

[2] Cook, DJ, Das, SK. How smart are our environments? An updated look at the state of the art. Journal Pervasive and Mobile Computing. 2007;3(2):53-73.

[3] Caragliu, A, Del Bo, C, Nijkamp, P. Smart Cities in Europe. Series Research Memoranda. University Amsterdam, Faculty of Economics, Business Administration and Econometrics. 2009;48:45-59.

[4] Gharavi, H, Ghafurian, R. Smart grid: the electric energy system of the future. Proceedings of the IEEE. 2011;99(6):917-921.

[5] Giffinger, R, Gudrun, H. Smart cities ranking: an effective instrument for the positioning of the cities?. ACE: Architecture, City and Environment. 2010;4(12):7-26.

[6] Airinei, D, Grama, A, Fotache, D, Georgescu, M, Munteanu, A, Dospinescu, O, Păvăloaia, VD, Popescul, D. The Use of Information and Communication Technologies in Organizations. Iași: Editura Universității Alexandru Ioan Cuza; 2013. 465 p.

[7] Borja, R, de la Pinta, JR, Álvarez, A, Maestre, JM. Integration of service robots in the smart home by means of UPnP: a surveillance robot case study. Robotics and Autonomous Systems. 2013;61(2):153-160.

[8] Neirotti, P, De Marco, A, Cagliano, AC, Mangano, G, Scorrano, F. Current trends in smart city initiatives: some stylised smart facts. Cities. 2014;38:25-36.

[9] Lee, JH, Hancock, MC, Hu, MC. Towards an effective framework for building smart cities: lessons from Seoul and San Francisco. Technological Forecasting and Social Change. 2014;89:80-99.

[10] Radu, LD. Green ICTs potential in emerging economies. Procedia Economics and Finance. 2014;15:430-436.

[11] Balena, P, Bonifazi, A, Mangialardi, G. Smart Communities Meet Urban Management: Harnessing the Potential of Open Data and Public/Private Partnerships through Innovative E-Governance Applications", Computational Science and Its Applications. Lecture Notes in Computer Science– ICCSA 2013. 2013;7974:528-540.

[12] Kitchin, R. Getting Smarter about Smart Cities: Improving Data Privacy and Data Security. Dublin, Ireland: Data Protection Unit, Department of the Taoiseach; 2016.

[13] Vermesan, O, Friess, P, editors. Internet of Things—From Research and Innovation to Market Deployment. Denmark: River Publisher; 2013.

[14] Camarinha-Matos, L, Afsarmanesh, H. Collaborative systems for smart environments: trends and challenges. Collaborative Systems for Smart Networked Environments, IFIP Series. 2014;434:3-14.

[15] Borgia, E. The Internet of Things vision: key features, application and open issues. Computer Communications. 2014;54(1):1-31.

[16] Popescul, D, Radu, LD. Data security in smart cities: challenges and solutions. Informatica Economică. 2016;20(1):29-39.

[17] Hossain, M, Fotouhi, M, Hasan, R. Towards an Analysis of Security Issues, Challenges, and Open Problems in the Internet of Things. In: Proc. IEEE 11th World Congress on Services (IEEE SERVICES 2015); June 27-July 2; New York. 2015. p. 21-28.

[18] Hewlett-Packard Enterprise. Internet of Things Research Study [Internet]. 2014. Available from: http://www8.hp.com/h20195/V2/GetPDF.aspx/4AA5-4759ENW.pdf [Accessed: 4 May 2016]

[19] Bertolucci, J. Big Data Drives the Smart Car [Internet]. 18 March 2014.Available from: http://www.informationweek.com/big-data/big-data-analytics/big-data-drives-the-smart-car/d/d-id/1127767 [Accessed: 5 May 2016]

[20] Maglaras, LA, Al-Bayatti, AH, He, Y, Wagner, I, Janicke, H. Social internet of vehicles for smart cities. Journal of Sensors and Actuators Networks. 2016;5(3).

[21] Muller, M. IoT Security: The Ugly Truth [Internet]. 25 September 2015.Available from: https://www.youtube.com/watch?v=j2qAkWDSDkg [Accessed: 10 May 2016]

[22] Agarwal, D. Testing Mobile Apps: Smartphones Sensors List [Internet]. 17 February 2016.Available from: https://testingmobileapps.wordpress.com/2016/02/17/smartphones-sensors-list/ [Accessed: 20 May 2016]

[23] Cilliers, L, Flowerday, S. Information security in a public safety, participatory crowd-sourcing smart city project. In: Proceedings of World Congress on Internet Security (WorldCIS-2014); 2014; New York, USA: Curran Associates, Inc. London, UK. p. 36-41.

[24] Loukas, G. Cyber-Physical Attacks: A Growing Invisible Threat. Elsevier; 2015.

[25] Owusu, E, Han, J, Das, S, Perrig, A, Zhang, J. ACCessory: password inference using accelerometers on smartphones. In: Proceedings of the Twelfth Workshop on Mobile Computing Systems & Applications; 2012; p. 9-15.

[26] Al-Haiqi, A, Ismail, M, Nordin, R. On the Best Sensor for Keystrokes Inference Attack on Android. In: The 4th International Conference on Electrical Engineering and Informatics (ICEEI 2013); 2013.

[27] Xu, Z, Bai, K, Zhu, S. Taplogger: inferring user inputs on smartphone touchscreens using on-board motion sensors. In: Proceedings of the fifth ACM conference on Security and Privacy in Wireless and Mobile Networks; 2012. p. 113-124.

[28] Michalevsky, Y, Boneh, D, Nakibly, G. Gyrophone: recognizing speech from gyroscope signals. In: Proceedings of the 23rd USENIX Security Symposium; 20-22 August; 2014.

[29] Appleby, T. Is NFC Still a Vulnerable Technology? [Internet]. 17 June 2013.Available from: https://securityintelligence.com/is-nfc-still-a-vulnerable-technology [Accessed: 15 April 2016]

[30] Gartner. Next year, smart cities will use 1.6 billion connected things [Internet]. 10 December 2015.Available from: https://www.helpnetsecurity.com/2015/12/10/next-year-smart-cities-will-use-16-billion-connected-things/ [Accessed: 14 May 2016]

[31] Millman, R. How Vulnerable are Smart Buildings to Cyber Hacks? [Internet]. 29 March 2016.Available from: http://www.ifsecglobal.com/how-vulnerable-are-smart-buildings-to-cyber-hacks [Accessed: 1 May 2016]

[32] Kaspersky Lab. Traffic Jams: Kaspersky Lab Discovers Security Issues in Smart Transport Monitoring System [Internet]. 18 April 2016.Available from: http://www.kaspersky.com/about/news/virus/2016/Traffic-Jams [Accessed: 18 May 2016]

[33] Popa, M, Cartas, C. OBD2 IoT device proof of concept for the insurance companies connected cars. In: Proceedings of the 15th International Conference on Informatics in Economy (IE 2016)—Education, Research & Business Technologies; 2-5 June; Cluj-Napoca, Romania. Cluj-Napoca, Romania: Bucharest University of Economic Studies Press; 2016. p.103-107.

[34] Rubin, A. All your Devices can be Hacked [Internet]. October 2011.Available from: https://www.ted.com/talks/avi_rubin_all_your_devices_can_be_hacked [Accessed: 10 May 2016].

[35] Sechel, S. Information insecurity: an assessment of the Romanian cyberspace. In: Proceedings of the 15th International Conference on Informatics in Economy (IE 2016)

—Education, Research & Business Technologies; 2-5 June; Cluj-Napoca, Romania. Cluj-Napoca, Romania: Bucharest University of Economic Studies Press; 2016. p. 314-320.

[36] European Union. Online Privacy [Internet]. [Updated: 11 April 2016]. Available from: https://ec.europa.eu/digital-single-market/node/39821 [Accessed: 13 May 2016].

[37] European Commission. Regulation (EU) 2016/679 of the European Parliament and of the Council of 27 April 2016 on the protection of natural persons with regard to the processing of personal data and on the free movement of such data [Internet]. 27 April 2016. Available from: http://eur-lex.europa.eu/legal-content/EN/TXT/?uri=CELEX%3A32016R0679 [Accessed: 20 May 2016].

5

Emerging Technologies for Renewable Energy Systems

Danilo Hernane Spatti and

Luisa Helena Bartocci Liboni

Abstract

Considering all aspects involving smart grid deployment, several subjects extrapolate the electrical sector. In the Brazilian scenario, it can be clear that power companies cannot support, by themselves, all steps for establishing renewable energy sources in smart grid systems. The technology demands are greater than what the electric sector can deliver. Such discussions about regulations of renewable energy sources are largely discussed in the society. The search for deploying eolic and solar generation with big energy farms are opposite to the smart grid and smart city renewable energy concept, which require decentralized actions. This chapter will show the concepts of eolic and solar energy sources specifically in the context of Smart Grids.

Keywords: renewable energy, solar and eolic energy, regulations, electrical vehicles

1. Solar and eolic energy in Brazil

With the publication of ANEEL Decree No. 440 in 2010, a working group was created to study the concept of smart grids. Such a group was intended to outline the various aspects, both technological and social, of the subject. The study had numerous findings of smart grid plans, both domestic and international, which point out that the Brazilian electric sector actually could not face, alone, the various technological and financial challenges for the implementation of smart grids [1].

In part, this is because the concept of intelligent networks very soon began to be linked to smart cities, which is composed of elements that go beyond the boundaries of energy generation, transmission, and distribution. Smart cities present a broader context of changing, because, in essence, cities must become more efficient in all its processes and services available to society.

Under this approach, in fact, the changing demands would be much broader than just those in the energy sector. This would require society mobilization to successfully carry out a transition from old models of electrical networks to the new smart grids [2].

Precisely, this broad background of discussions on smart grids and smart cities highlights that the sources of renewable energy are an agenda that deserves much attention. The Brazilian energy matrix is considered clean, but some factors such as cost and implementation time of new hydropower plants have favored the emergence, in recent years, of thermal plants. Today, thermal plants account for almost 29% of the installed capacity of the country's energy generation. Used as alternatives to meet emergency demands, the thermal plants take advantage of its speed of deployment, fine-grained control of generation, and flexibility in the fuel that is used. In addition to its construction being expensive, such plants have a strong social appeal as they do not qualify as a clean source of power generation, requiring fossil fuel to burn for its operation.

It is in this context that an increasing need for alternative energy sources is predominant. The search for alternative energy sources is also a worldwide concern, as can be seen in **Table 1**.

An accelerated global growth can be seen in the deployment of the generation capacity of alternative energy sources from 2010, led by China and the United States. These nations, which notably possesses the greatest industrialization level for the period analyzed, are in the power generation forefront, whether or not alternative sources meet their incremental internal demands.

Such statements only reinforce the commitment of many countries to seek alternative energy sources, which are more efficient, cheaper, and meet the expectations of society for lower environmental impacts.

Country	2007	2008	2009	2010	2011	Δ% 2010/2011	Total %
China	8.2	15.0	19.3	36.4	73.7	102.3	18.8
USA	30.9	39.4	49.4	54.7	64.3	17.5	16.4
Germany	30.9	34.3	41.6	51.3	61.3	19.4	15.6
Spain	16.3	20.8	23.7	26.3	27.0	2.9	6.9
Italy	4.7	6.1	8.6	12.2	23.2	89.7	5.9
India	9.3	11.8	13.2	15.7	20.0	27.8	5.1
Brazil	6.6	7.4	6.7	8.8	12.4	41.2	3.2
France	3.6	5.0	6.3	8.5	11.5	34.5	2.9
UK	4.3	5.2	6.4	7.7	10.8	39.7	2.8
Japan	5.5	5.9	6.7	8.0	9.5	19.3	2.4
Others	40.2	46.5	55.6	63.8	78.7	23.4	20.1

Source: Adapted from reference [3].

Table 1. Installed capacity of renewable energy in the world (GW).

In Brazil, this concern resulted in the growth of the installed capacity of renewable energies of around 41% between 2010 and 2011, following the trend of other nations [3].

Among the several ways currently available for generating power from renewable sources, we can highlight two ways, which are the most cited by the technical-scientific community: solar energy and wind energy. **Figure 1** shows a representation of the installed generation capacity of solar and wind energy in Brazil.

According to reference [3], the evolution of the installed generation capacity of wind and solar energy in Brazil has been growing at different levels since 2007. While wind farms accounted for 1.7% of the total installed capacity of the country in 2013, solar plants had lower expressibility, up to 0.0% of the total installed capacity. In 2011, about 2200 MW corresponded to generation through wind; however, only 5 MW were declared as the installed capacity of solar power plants.

Brazil has an innate ability to generate solar energy because of its location in a tropical area with a high incidence of useful radiation throughout the year. Notably, one of the major obstacles to implementing this type of generation center is the need for large footprint.

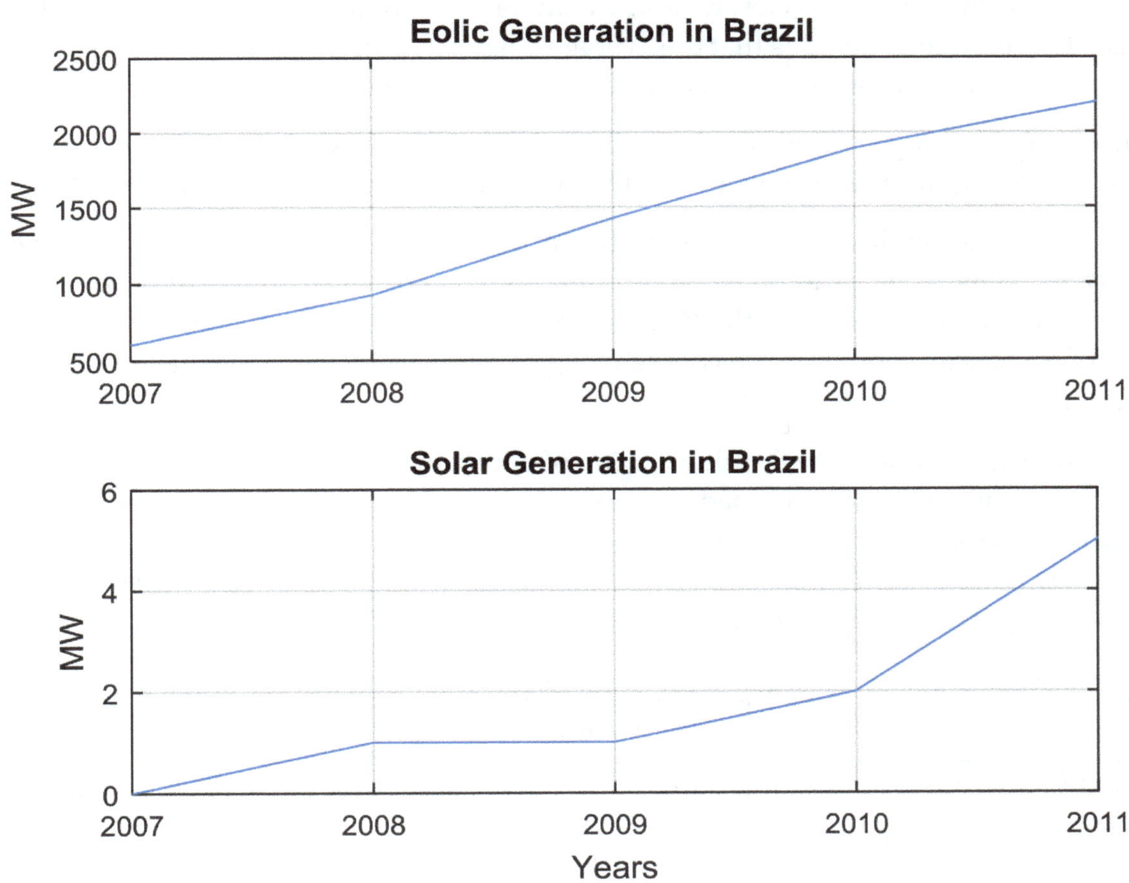

Figure 1. Installed capacity of eolic and solar generation in Brazil. *Source:* Adapted from reference [3].

This is partly because technologies involving solar capture and conversion into electricity through photovoltaic cells are still being constantly researched to find more efficient ways to harness sunlight.

Recently, to increase the installed capacity of solar power generation, the Energy Research Company (EPE) enabled 11,261 MW of 341 projects from solar PV for auction. A large number of qualified projects will surely promote competitiveness in the auction, which is also needed for research and investment seeking more efficient ways of establishing these plants.

The volume of energy is significant and comparable to large hydroelectric projects like the Belo Monte Dam. The state with the largest number of projects is Bahia, which could add 3998 MW to the National Interconnected System (SIN). Piauí state had enabled 61 projects (2077 MW) and São Paulo state 37 (1293 MW) [4].

Regarding wind power generation, Brazil's installed wind power capacity is concentrated in three main regions: the Northeast Coast, being Rio Grande do Norte, Ceará, and Piauí states; Bahia, which also comprises the interior of Pernambuco state; and Rio Grande do Sul, which accounts for more than 90% of the entire installed wind farms in the country. According to reference [5], the average statistics regarding these generation parks are summarized in **Table 2**.

The energy generated on the coverage area of the turbines is a very important characteristic because the catchment area is closely linked with the wind regime of the region. Another factor that should be considered is that the average height of generators is above 80 m in height, which provides monumental landscapes from the wind farms and costly installation.

The need for such high turbines comes from deep investigations concerning the system of winds. If such requirement is not attended the wind potential is not entirely availed.

The wind behavior in Brazil has been studied since the 1970s, showing a prospective scenario for this type of energy generation in the country. The "Atlas of the potential of Brazilian Wind" [6] pointed out that by the end of the 1980s, with measurements up to 10 m tall, trends in wind speeds of 5 and 6 m/s could be considered along coastal and inland areas that have favorable topography and roughness.

In the next decades, following the global trend of deploying wind power on a large scale, new studies have emerged in Brazil, made by several companies in the electricity sector promoting feasibility reports of towers of 30–50 m in height. The government of Ceará state in 2001, for example, published an Atlas of the Wind power generation potential of the state, pointing generation capabilities with average wind speeds of around 7 m/s using towers of 70 m.

Basins	Energy index (W/m^2)	Production index (kWavg/machine)	Wind generators	
			Average power (kW)	Average height (m)
Northeast Coast	345.4	860.9	1934	87.6
Bahia	471.3	812.1	1715	83.5
Rio Grande do Sul	329.3	660.2	2000	82.0

Table 2. Main eolic regions in Brazil.

Therefore, it is possible to affirm that Brazil has great potential for deploying solar and wind power stations, especially large-scale projects, which are already being conducted. The high importance of these projects is, in part, justified by two reasons.

The first reason lies in the fact that Brazilian system (SIN) is interconnected so that each region can have its power generation potential leveraged in the energy matrix, being it hydro, solar, wind, etc., more efficiently. The energy generated is then transported to the consumer centers. This is one of the practices most used by the Brazilian electric sector currently focusing on the construction of large plants that can make the most of the energy potential of the region in which they operate.

The second reason lies in the fact that consumption centers are far from generating centers, especially in the Southeast region, which is the major consumer of energy in the country. In reference [6], the authors point out that the average distance between generation and consumer centers ranges from 500 to 1000 km, reinforcing that the creation of large power plants is economically justified, since they are installed where there is a better climate, relief, watershed, etc.

This Brazilian model of large power plants has been debated, in various social forums. This subject still needs further discussion, since several sectors have proved incompatible with such model.

2. Solar and eolic energy in the microgrid context

The various factors mentioned in the previous section and that punctuate the Brazilian policy of creating large power plants are opposed to the concept of intelligent networks with the presence of renewable energy.

That is because the context sought by smart cities is precisely that of decentralized energy production, which would be a more appropriate setting for power distribution with distributed generation capacity, where customers could enter the energy mix as well as generators.

ANEEL, through Normative Resolution No. 482 [7], established the general conditions for the micro- and minigeneration access to distributed power distribution systems and electrical power compensation system. This was a response to wishes from the scientific community to the theme of distributed generation, which for a long time was already promoting articles in conferences and journals.

Since 2006, it can be seen in the literature alignment of research involving smart grid topics where distributed generation is also discussed. In this context, small generating units were given the name microgrid. A technological step was necessary to explore more efficiently the system that is now available, such as powerful signal processing tools, data transmission, storage, and knowledge discovery in databases [8].

The distinction in Brazil's microgeneration and minigeneration standardized by ANEEL is used for the power generating unit, as can be found in reference [7]. Microgenerators are the ones with less than 100 MW of installed power and minigenerators are the one that have an installed power between 100 and 1000 MW.

In both cases, renewable energy generation may be used alone or along with high energy efficiency systems, such as hydro, solar, wind, biomass, or cogeneration, connected to the distribution network through consumer units. Unlike what some European countries envision, where in fact consumers may be remunerated by the sale of electricity to companies, in Brazil, consumers make part of the established power compensation system.

In this case, the active energy injected by a consumer unit with micro- or distributed minigeneration is transferred to the local distributor and later compensated with energy consumption. Therefore, a consumer with small generating units, such as photovoltaic panels and wind solar, uses the energy generated to compensate their energy bill. The energy excess is converted into abatement credits for future accounts or current accounts of another consumer unit of the same client. Credits are valid for 36 months [9].

Among the many developments promoted by Normative Resolution 482/2012, it may be noted that it is not established, technically or economically, by ANEEL, the particular characteristics of such generators. It is the consumer's responsibility to evaluate what would be the best conditions of cost/benefit for use in micro- and minigeneration.

The dialogue between customers and distribution companies is also stimulated as companies cannot refuse to perform the connection of the consumer's generating units, as ANEEL understands that the companies are responsible for ensuring reliable energy utility.

Thus, the indicated benefits, such as postponement of investments, reduced load, low environmental impact, among others, shall only be advantageous for both consumers interested in generating energy, and distributors interested in finding technical and economically viable solutions.

The authors of reference [9] state that complementary to Resolution 482 are Module 3 PRODIST [10] and the Resolution 414/2010 [11]. These documents are made and summarized for customer units connected at low voltage (group B). If the energy injected into the network exceeds the consumption, it will result in payment for the cost of availability: 30 kWh (single), 50 kWh (two-phase), or 100 kWh (three-phase).

For customers connected to a high voltage (group A), the invoice energy portion will be compensated, but the share of the demand will be billed normally. According to ANEEL's data, currently, there are 1466 micro- and mini-generation registered agents associated with Resolution No. 482/2012.

The large-scale use of photovoltaic panels, photothermal, and small wind turbines for homes and businesses certainly would be characterized as a new power generation paradigm, different from the use in the so-called generation farms.

It is precisely these actions involving smart grid initiatives that engage the distributed generation system in the world, where each consumer would be a key element responsible for supplying energy in a mobile way. Although the scientific community focus efforts in managing large renewable power plants, which are extremely costly for deployment and management, enterprises that pursue the compatibility of small generating units for supplying loads in distribution networks are of great value [12].

The compensation system proposed by ANEEL for mini- and microgeneration is setup for small applications, following the global trend of distributed generation, by fostering the rise of small distributed generation with great diffusion in homes and businesses. In this new paradigm, where each consumer could own their microgenerator and make better use of renewable energy sources available in their home, some initiatives have gained prominence.

2.1. Heliothermic energy

According to the Society of the Sun [13], which supports low-cost solar heaters projects, it is possible to build heliothermic systems up to half the price of conventional systems. This organization provides free step-by-step execution of the project represented in **Figure 2**.

Simply replacing the electric shower by heliothermic plates, for example, could represent a reduction of 3–9 kW in the loading of homes, which generates savings in the energy bill and also in the protection circuit, which could use conductors and circuit breakers with lower nominal currents.

The materials to be used in the construction of the solar heater needs to be inexpensive, and easily accessible. The system takes advantage of the heat exchange surface of the PVC, painted with black ink, and a capillary water system also made from a PVC tubing. When in contact with the heated surface, water exchange heat, becoming warmer, being stored in a reservoir.

This system exploits the principle that a water tank itself can become a storage element for thermal energy when it has a large amount of heated water. Then, the water is dispensed using special containers. The thermal inertia of the large volume of water and the hot water flux on top of the tank ensures maximum system efficiency. A graphical representation of the system can be seen in **Figure 3**.

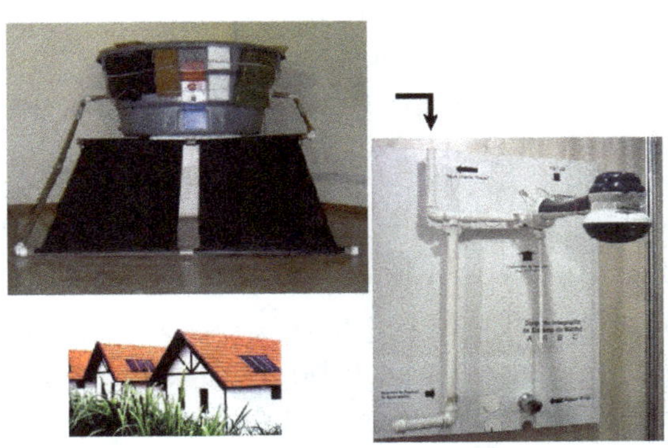

Figure 2. ASBC project. *Source:* Adapted from reference [13].

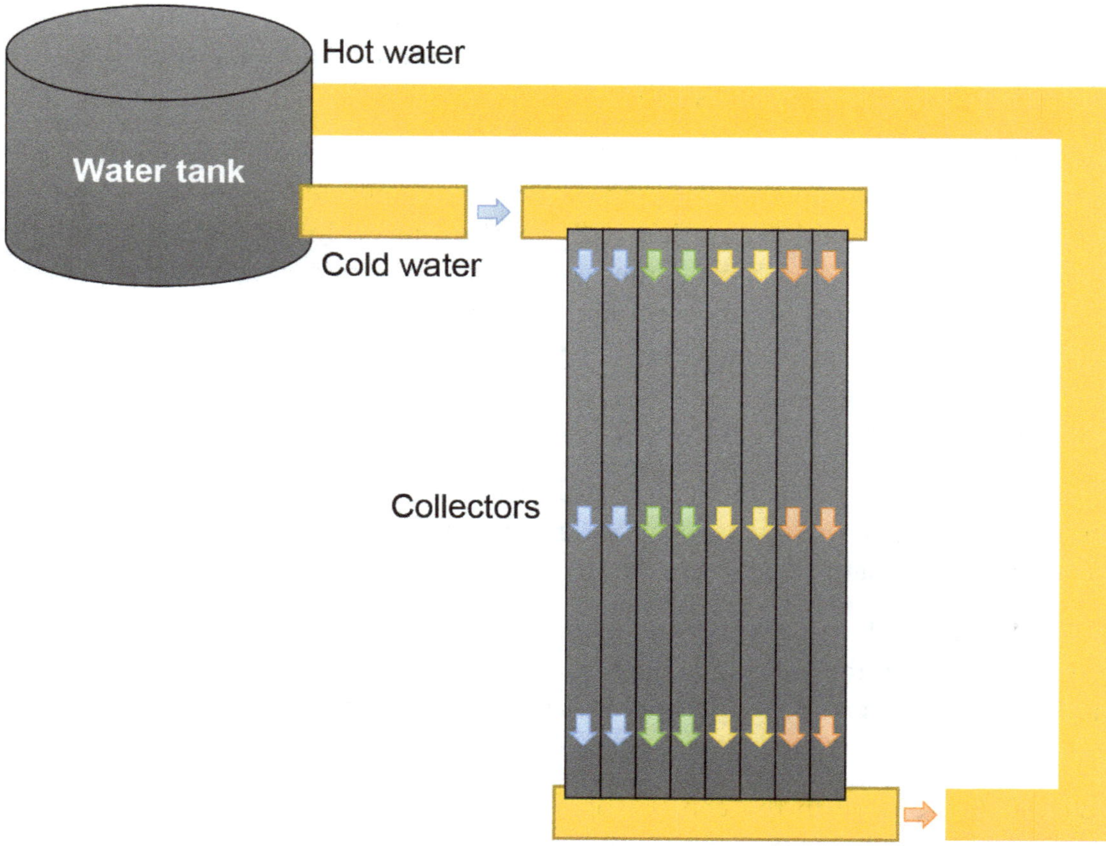

Figure 3. Low-cost solar collector operation.

According to the creators of the project, for a 300 L tank, the area of the collector is about 1.5—2.5 m^2, with a total weight between 15 and 40 kg/m^2 depending on the type. The system slope is ideally set up. However, it still needs a slope adjustment in winter. The roof slope should be close to the local latitude, or up to 10 degrees more. The initiative of the Sunshine Society offers several other projects to avail the Brazilian heliothermic potential at a low cost. The production scale of this projects certainly would provide competitive cost on conventional system, which is still not intensively used due to the high investments and longer return time.

2.2. Photovoltaic energy

Photovoltaic cells still have little market penetration, which is completely contradictory to the potential generating capacity of this system. According to reference [14], considering only the area of the Earth where the deployment of solar power plants would be possible, the energy generated would be around 1011 GWh/year.

As sunlight is composed of a directly incident part and another part that is diffusely incident, the efficient use of this form of energy requires collectors that align with the sun. Otherwise, only a portion of the generation capacity will be available. In Brazil, according to reference [14], the largest global radiation value occurs in Bahia state and the lowest in Santa Catarina state. However, solar radiation values in any region of Brazil are higher than in many countries where the rate of use of this form of energy is higher than in Brazil.

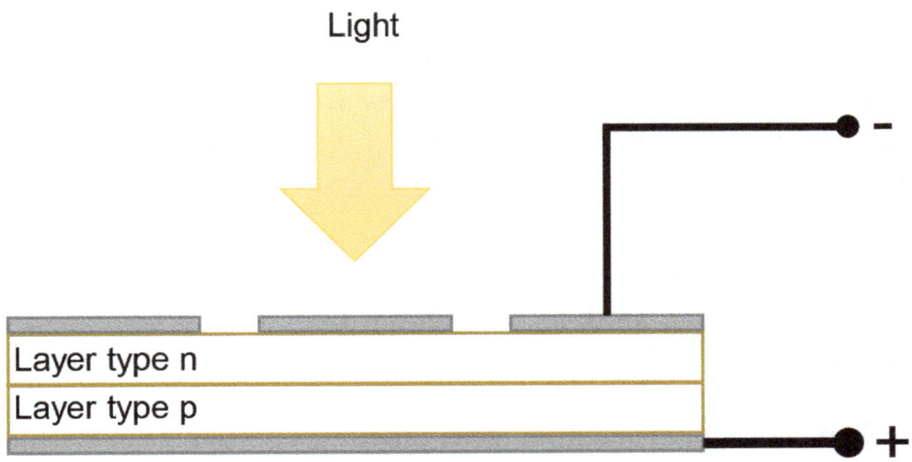

Figure 4. Photovoltaic panel representation.

In **Figure 4**, a schematic representation of a photovoltaic collector plate is shown, where you can check the distribution of type p and n layers and the metal arrangement responsible for the voltage supplied to the circuit.

The semiconductor substrate material, when bombarded by sunlight, generates a voltage which is available in the metallic terminals. To increase generation capacity combinations of these devices are used, since each has an average energy efficiency of less than 15%, that is, 85% of the radiation incident on the light collector is not transformed into energy. The lifecycle of these collectors is also a contributing factor for its high cost to consumers, since it is exposed to the sun, and its low efficiency tends to become even smaller over time when compared to other forms of power generation.

2.3. Wind turbines

Wind turbines take advantage of the kinetic energy of wind captured within the coverage area of the rotor blades to generate electrical power. By reaching the wind turbines, the wind tends to decrease its speed; however, by forming a helical vortex, speed is recovered, since a minimum distance between turbines is required for minimizing loss of performance [6].

A safe working distance is considered to be five times the rotor diameter between alongside turbines and 10 times the diameter between downstream turbines. The rotor speed is inversely proportional to the blade diameter, and speeds of the order of 15–30 rpm are considered slow for practical implementation criteria. This type of practice rule ensures that the blades are visible and avoidable by birds in flight. Another aspect to be considered is that the higher the speed, the higher the rotor noise level.

Initial rates for availing wind lie between 2.5 and 3.0 m/s. Due to the high inertia of the rotor in situations where wind speeds are between 12.0 and 15.0 m/s, it is necessary to activate the power-limiting system of the machine. In situations where the wind regime reaches higher levels, new studies on the feasibility of the deployment become necessary, since it can undergo large efforts from the structure of the machine.

The IDEAL Institute [15], which promotes projects for the development of renewable energy in Latin America, produced a guide with the support of ANEEL and several other entities, to promote the dissemination of information to those who want to install a wind turbine at home. This guide clarifies some findings regarding wind turbine installations in large generation parks, targeting correct orientation for consumers who can benefit from the tax incentive introduced by ANEEL.

According to the guide, small wind systems are closer to the ground than large wind turbines, and analysis of the land around the facility is of great importance for properly utilizing the potential of the winds. As pointed out by reference [6], the wind regime has an increase in speed with greater heights, and at lower heights speeds are mitigated by a multitude of factors. Typically, micro and mini wind generators need to be installed on high towers or top of buildings. **Figure 5** shows some safety distances given by the IDEAL Institute's guide.

According to the Atlas of the Brazilian Eolic Potential [6] along with the IDEAL Institute guide, it can be seen that wind turbine installations for the purpose of micro-/minigeneration has an average wind speed of 6–7 m/s according to the minimum height specification, with good generation capacity. However, as there are a plethora of models available, one should consider the noise that each model produces, it is also required to simulate noise level in the neighborhood. Large wind turbines of wind farms have acceptable noise level of 300 m, so this is a very important factor in choosing the model.

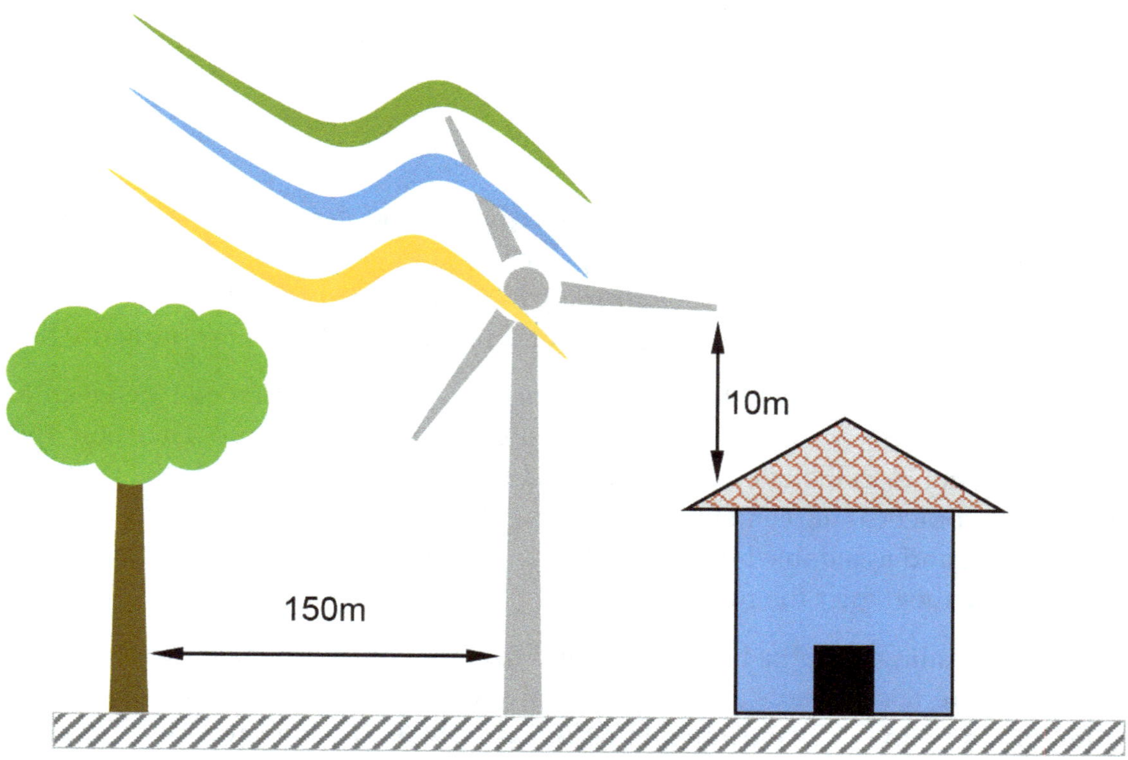

Figure 5. Minimal distances between wind turbines. *Source:* Adapted from reference [15].

As the efficiency of this type of power supply is relatively high, aspects such as power generator size, rotor size, etc., must be thoroughly investigated by consumers so that he can obtain a good cost and benefit relation.

Another important factor concerns the protection of equipment and people. As mentioned above, the speed control system is crucial for safe operation for humans, animals, and properties.

As turbines are installed on high towers, consumers need to keep in mind that a proper grounding system and possibly a new grounding should be considered.

3. Electrical vehicles in the microgrid context

It is estimated that there are approximately 900 million transportation vehicles circling the planet, where 90% of these depend entirely on fossil fuels and internal combustion engines, despite these vehicles being declared obsolete for decades due to their high greenhouse gas emissions and their low efficiency. Effectively no more than 30% of the energy contained in the fuel is transformed into vehicle's workforce. The emission of gasses such as carbon dioxide, nitrous oxide, and chlorofluorocarbons from combustion engines has generated debates in numerous international discussion forums, where actions for minimizing the effects of global warming were in focus.

Therefore, rapid growth has been seen in the electric vehicles industry, whose main purpose is to replace traditional vehicle systems and their high carbon emissions. The deployment of smart grids come at the forefront of a promising scenario to promote change for electric vehicles. The main reason for this prospective scenario is the extrapolation of the concept of smart cities, which brings more efficient, environmentally friendly, and affordable products and services to society [16].

According to reference [17], a major challenge involving the spread of electric vehicles is the strategic distribution of charging points. This topic is certainly what concerns consumers the most when there is the possibility of only charging their vehicle at home. To this end, it is important to gradually establish a broader availability of technologies, so that the user gets used to the idea of abandoning its vehicle using fossil fuel.

Hybrid electric vehicles are those that use a combustion engine together with electric mothers. This was the first technology presented, which started preparing consumers for electric vehicles. Hybrid electric vehicles can effectively be plugged in an outlet to charge their battery banks, in addition to working with combustion engines.

The batteries, in this case, can provide extra mileage range when fuel ends. Vehicles battery-powered only and vehicles with hydrogen fuel cells are the evolution of the hybrid technology, in which combustion engines would no longer be necessary, reducing the greenhouse gas emission.

There are currently several types of batteries in the market, with different features for different application and that may or may not be suitable for use in electric vehicles. In reference [18], the authors list the types of batteries that were used in their study for managing and dimensioning energy supplies for electric vehicles, which can be highlighted as follows:

- Lead-Acid Battery: it is the battery pack used in motor vehicles. It has good reliability when operated with low levels of loading and unloading, also being used in different versions of UPS batteries. To use it in electric vehicles, adjustments are necessary to replace the liquid chemical element with gel to prevent leakage since electric motors require a higher level of discharge.
- Nickel Cadmium Battery: is considered the most efficient batteries to operate at high discharge environments and can provide very high levels of current. Its operation in electric vehicles, however, needs further investigation since the disposal of cadmium can be extremely harmful.
- Lithium Ion Battery: presents high efficiency and low cost as its life is determined by charge/discharge cycles. It works with a very high discharge level, promoting high torque for the electric motors.
- Lithium Ion Battery: this is the most modern lithium batteries in the market. They work with a very high level of discharge; it can also support bigger current regime, which reduces the recharge time.

Supercapacitors are another energy storage source for electric vehicles. Supercapacitors are elements capable of operating with loading and unloading of large amounts of energy, especially suitable for driving large electric motors. Several studies have promoted the use of these elements in electric vehicles for public transportation, which requires larger engines and lower recharge times. Moreover, batteries can be charged at the boarding terminal [18].

The use of supercapacitors and batteries consists of a combination of energy sources technologies capable of supporting both long-haul routes when moving extremely heavy vehicles, a situation still unsupported by current batteries performing alone

In the context of smart grids, considering a heterogeneous environment of electric vehicles having the most diverse types of batteries and supercapacitors, a management of strategic points of recharge is necessary. Mass transit vehicles, for example, using supercapacitors require a high voltage source for rapid charging, while vehicles with battery sources require lower voltages and currents.

You can group the types of loading according to the charging time [19]:

- Slow charging: usually applies a charging current of about 1/10 of the current capacity of the battery. It does not heat and deteriorate the internal elements of the battery. This procedure popularized lithium batteries and nickel.
- Fast charging: uses a recharging current equal to or greater than the battery charging capacity. This type of operation provides plenty of stress to the battery, decreasing its life if it is not designed for this purpose. Lithium ion batteries have shown good response with very high levels of recharging, performing a quick battery recharge to operating levels.

Recharge time will be a determining factor in the study and expansion of recharge points in smart grids. Power agents recharging points shall also take into account the level of hybridization of the vehicle, driving patterns, the types of batteries available, and also seasonal issues. Electric vehicles will only be successful if they indeed resemble the vehicles consumers use today, i.e., with multimedia stations, air conditioning, heaters, and other electrical optional items. Some items depend on environmental conditions.

On a very hot day, for example, the distributor must be prepared for a larger load at the recharging points due to the intense use of air conditioning by electric vehicles. If the traffic is jammed on a hot day, this can worsen the situation because vehicles are consuming more energy without moving. Such a situation could overload the rapid charging points for vehicles [17].

When considering a conventional power network, which could support the loading of all vehicles during the night, it could result in a contribution of investments in new generating plants far superior to what would be gained with a massive penetration of distributed generation. However, in the context of smart grids, the shipments could be fired in cycles, ensuring maximum efficiency of power during the charging process. Another possibility is differentiated charging, which causes the consumer to adhere to their vehicle loading programs with lower charging times.

Other clusters of discussions are energy storage for electric vehicles during the braking process and recharging using photovoltaic panels. The energy stored while braking could be used via a kinetic retrieval device from the vehicle, such as overtaking or recharging processes of other devices. With the capability of detecting such vehicles connected to intelligent networks, as soon as their batteries were loaded, photovoltaic panels on the roof of vehicles could continue to provide power to the grid, functioning as a distributed solar generator.

In reference [17], a synthesis of several case studies involving the use of electric vehicles in many different types of smart grids is displayed. Bellow, we point out some noteworthy aspects raised in these case studies:

- Overload in transformers: from several simulations using different transformers, it can be seen that a 25 kVA transformer of were overhead when using a quick charging condition of the vehicle in peak hours and operating on the edge outside the peak hours. When slow recharging, the transformer remained within its operating limits. Using load scheduling of vehicles with the smart grid transformer operating limits was also maintained.
- Integration with the network: A study promoted in Belgium showed that the integration of electric vehicles in the distribution network had negative impacts of power losses and also of voltage variations. In conclusion, recharging should be controlled via a centralized operator.
- Impacts on the distribution systems: Studies have indicated that there may be increased temperature of transformers due to increased load, high harmonic levels from the switching power supplies needed to recharge the vehicles, and bushing wear of the transformer.

4. Energy storage in smart grids

Energy storage is a pretty major topic in forums involving various areas of engineering. In the context of smart grid, this topic also deserves special attention, since in a scenario with a strong presence of distributed generation, the idea that every consumer could generate energy for their demands and inject the surplus into the network becomes feasible.

However, a renewable energy system has unique characteristics and how to store any extra energy generated should be analyzed with caution and targeting efficiency.

This debate occurs for solar photovoltaics collectors. With such low efficiency in the generation, the potential of these collectors needs to be leveraged. As during the night, collectors do not generate power, if there is excess produced during the day, it would be interesting to store. In the context of micro- and minigenerators proposed by ANEEL, the client has a second option of injecting the surplus power into the network.

Some designs of microgrid systems propose the use of the entire roof of homes as a source of collecting solar radiation. Part of the energy is intended for heating water and the other to generate electricity. The presence of a wind turbine would result in all three systems functioning, which would result in a much higher production capacity for the consumer demand.

Even with the possibility of a surplus sale, certainly, the consumer should contemplate the possibility that at any contingency of the distribution system, their demands could be met through a stored portion of its energy generated.

The solar thermal energy would be stored taking into account the number of people who would use this energy as well as the minimum time solar radiation is needed to reach the desired temperature for water heating. It is prudent also to consider the situations where the level of sunlight would fall to a minimum, especially in winter, protruding compatible reservoirs and collectors with a strategic reserve of water for at least 3 days without sun.

According to reference [6], wind turbines have higher energy generation during the daytime. Photovoltaic generators only produce energy during the day, so, for both these types of renewable energy, it is important to implement systems for energy storage.

In fact, the nonpredictability of natural resources used to operate the generators shows an interesting issue. A solution is pointed out by using energy storage devices at times of low demand to regularize consumption in times of tip [20].

The combined operation of various renewable sources of energy being used in smart grids are presented in various sources of technical and scientific literature [21].

Energy sources such as solar and wind usually are used with a mandatory energy storage system, just to meet the demand in times where wind or sunlight are not offering the required energy [22]. Such systems have associated environmental impact since they use lead-acid battery banks [23].

Recent literature points out that this factor can be solved with the development of new energy storage technology, such as [24–28]:

- Clean batteries: are batteries that are based on the generation of energy from organic iterations.
- Fuel cells: use hydrogen electron exchange mechanism in isolated compartments.
- Compressed air: in times of low demand, generators utilize the surplus energy to power air compressor coupled to huge reservoirs, which then have the air released controllably driving a turbine system.
- Accumulators: devices such as supercapacitors may be used for accumulation and discharging of large amounts of energy.

5. Renewable energy systems integration in a smart grid

The integration of various renewable energy systems that can be implemented in the context of smart grids foments research involving a combination of tools. The justification for this research empowering is that, in the smart grid environment, a plethora of information about the system is available, providing an avalanche of data that must be properly addressed.

As mentioned before, at first, available information is not observable requiring the establishment of rules that relate the available informational entities. Some combination of tools gives rise to the so-called expert systems.

From a conceptual point of view, expert systems are those particularly distinguishable in their field, bringing together a multitude of techniques to analyze data and that have a high level of interoperability, and are robust and capable of providing efficient results even with data containing uncertainties.

When in operation, a large amount of information of all consumers would be available, such as billing, protection, losses, power quality, among others, since data communication is an essential part of the design of any smart grid.

Resolution 482 from ANEEL dictates only the maximum sizes of generators, leaving it free for consumer and distributors to define technical aspects used to integrate the mini- or microgenerating unit to the system. The equipment market being highly heterogeneous, it is evident that defined procedures and requirements need to be established to result in customer satisfaction and quality service from the distributor.

Considering the alternative sources of power generation various equipment are recommended, namely, frequency inverters, which are needed to ensure correct levels of voltage and frequency for generators operation in the distribution network and energy storage. As already discussed, the most usual storage method used in micro- and minigeneration are made by banks of batteries, formed by associations of lead-acid batteries packed in compartments with monitored temperature and gas emissions.

Notably known for large harmonic content they produce, the inverters can be configured as technological villains, reversing clean energy gains, since these devices were polluting the network with waveforms not purely sinusoidal.

In addition to these, various technological challenges for the integration of distributed generation systems in the context of smart grids are mentioned below:

- Power injection: The entry of the generators must be synchronized on the network to not cause instability problems.
- Protection: Making sure that the generators will not contribute any system failures, such as food shortages.

Author details

Danilo Hernane Spatti[1*] and Luisa Helena Bartocci Liboni[2]

*Address all correspondence to: danilospatti@utfpr.edu.br

1 Federal Technological University of Paraná, Curitiba, Brazil

2 Federal Institute of Education, Science and Technology of São Paulo, University of São Paulo, São Paulo, Brazil

References

[1] ANEEL, "Portaria N° 440". Criação de Grupo de Trabalho com o objetivo de analisar e identificar ações necessárias para subsidiar o estabelecimento de políticas públicas para a implantação de um Programa Brasileiro de Rede Elétrica Inteligente - "Smart Grid". Abril de 2010.

[2] Grupo de Trabalho de Redes Elétricas Inteligentes, "Smart Grid". Relatório, Ministério de Minas e Energia, 2011.

[3] EPE, Empresa de Pesquisa Energética (Brasil). Anuário Estatístico de Energia Elétrica 2014 – Rio de Janeiro: EPE, 2014.

[4] EPE, Empresa de Pesquisa Energética (Brasil). Habilitação de Projetos Para Leilão de Energia Solar – Rio de Janeiro: EPE, Agosto de 2015.

[5] EPE, Empresa de Pesquisa Energética (Brasil). Boletim Trimestral da Energia Eólica – Rio de Janeiro: EPE, Agosto de 2015.

[6] Ministério de Minas e Energia, Atlas do Potencial Eólico Brasileiro, Brasília, 2001.

[7] ANEEL, Resolução Normativa N° 482, Brasília, 2012.

[8] Katiraei, F., Iravani, M.R., Power Management Strategies for a Microgrid With Multiple Distributed Generation Units. IEEE Transactions on Power Systems, vol. 21, n. 4, pp. 1821–1831, 2006.

[9] ANEEL, Perguntas e Respostas sobre a aplicação da Resolução Normativa n° 482/2012, Brasília, 2012.

[10] ANEEL, PRODIST Módulo 3 – Acesso ao Sistema de Distribuição. Brasília, 2012.

[11] ANEEL, Resolução Normativa N° 414, Brasília, 2010.

[12] ANEEL, Micro e Minigeração Distribuída. Cadernos Temáticos ANEEL, Brasília, 2014.

[13] Sociedade do Sol, "MANUAL DE MANUFATURA E INSTALAÇÃO EXPERIMENTAL DO – ASBC – AQUECEDOR SOLAR DE BAIXO CUSTO". 2009.

[14] Torres, R.C., "Energia solar fotovoltaica como fonte alternativa de geração de energia elétrica em edificações residenciais". Dissertação de Mestrado apresentada à Escola de Engenharia de São Carlos como parte dos requisitos para obtenção do título de Mestre em Ciências – Pós-Graduação em Engenharia Mecânica, 2012.

[15] IDEAL – Instituto Para o Desenvolvimento de Energias Alternativas na América Latina, Guia de Micro Geradores Eólicos, São Paulo, 2014.

[16] Osorio, V.A.G., "Carregamento Ótimo de Veículos Elétricos Considerando as Restrições das Redes de Distribuição de Média Tensão". Dissertação apresentada à Faculdade de Engenharia – UNESP – Campus de Ilha Solteira, como parte dos requisitos para obtenção do título de Mestre em Engenharia Elétrica (Área de Conhecimento: Automação), 2013.

[17] Pereira Jr, L.C., "A Interação Entre Geradores Solares. Fotovoltaicos E Veículos Elétricos Conectados À Rede Elétrica Pública". Dissertação submetida à Universidade Federal de Santa Catarina com requisito parcial exigido pelo Programa de Pós- Graduação em Engenharia Civil – PPGEC, para a obtenção do título de MESTRE em Engenharia Civil, 2011.

[18] Lopes, J., "Metodologias De Dimensionamento E De Gestão De Fontes De Energia Para Veículos Elétricos". Tese de Doutorado apresentada à Faculdade de Engenharia Elétrica e de Computação da Universidade Estadual de Campinas para obtenção do título de Doutora em Engenharia Elétrica. Área de concentração: Automação, 2012.

[19] Vilhena, N.H., "Análise Do Carregamento Do Carro Elétrico No Contexto Smart Grid". Trabalho de Conclusão de Curso submetido ao Colegiado de Engenharia Elétrica para a obtenção do Grau de Engenheira Eletricista.

[20] Castro, R., Lyra Filiho, C., "Um método de suporte a decisões sobre investimento e comercialização de energia elétrica no Brasil". SBA Controle & Automação, vol. 16, no. 4, pp. 478–494, 2005.

[21] Shim, J.W., Cho, Y., Kim, S.J., Min, S.W., Hur, K., "Synergistic Control of SMES and Battery Energy Storage for Enabling Dispatchability of Renewable Energy Sources". IEEE Transactions on Applied Superconductivity, vol. 23, no. 3, pp. 2062–2065, 2013.

[22] Xu, Y., Singh, C., "Power System Reliability Impact of Energy Storage Integration With Intelligent Operation Strategy". IEEE Transactions on Smart Grid, vol.5, no. 2, pp. 1129–1137, 2014.

[23] Soares, M. P., Marcato, A.L.M., "Otimização linear seqüencial para cálculo de energia firme das usinas hidrelétricas do sistema interligado nacional". XVI Congresso Brasileiro de Automática (CBA), Salvador, pp. 1962–1967, 2006.

[24] Hreinsson, E.B., Barroso, L.A., "Defining optimal production capacity in a purely hydro-electric power system". IEEE International Conference on Electric Utility Deregulation Restructuring and Power Technologies, Hong Kong, vol. 1, pp. 178–183, 2004.

[25] Faria, E., Barroso, L. A., Kelman, R., Granville, S., Pereira, M. V., Iliadis, N., "Allocation of firm-energy rights among hydro agents using cooperative game theory: an aumann-shapley approach." Operation Research Models and Methods in the Energy Sector, Portugal, 2006.

[26] Sahyani, R.A., Oliveira, M.A.G, "Externalidades Da Geração De Energia Com Fontes Convencionais E Renováveis". Congresso Brasileiro de Planejamento Energético, 16 p., 2009.

[27] Converse, A.O., "Seasonal Energy Storage in a Renewable Energy System". Proceedings of the IEEE, vol. 100, no. 2, pp. 401–409, 2012.

[28] Grbovic, P. J., Ultra-Capacitors in Power Conversion Systems – Applications, Analysis and Design from Theory to Practice. John Wiley & Sons Ltd, Chichester, United Kingdom, 336 p., 2014.

6

Sensing Human Activity for Smart Cities' Mobility Management

Ivana Semanjski and Sidharta Gautama

Abstract

Knowledge about human mobility patterns is the key element towards efficient mobility management. Traditionally, these data are collected by paper/phone household surveys or travel diaries and serve as input for transportation planning models. In this chapter, we report on current state-of-the-art techniques for sensing human activity and report on their applicability for smart city mobility management purposes. We particularly focus on the use of location-enabled devices and their potential towards replacing traditional data collection approaches. Furthermore, to illustrate applicability of smartphones as ubiquitous sensing devices we report on the use of Routecoach application that was used for mobility data collection in the city of Leuven, Belgium. We provide insights into lessons learned, ways in which collected data were used by different stakeholders, and identify existing gaps and future research needs in this field.

Keywords: smart cities, travel behavior, travel patterns, data collection, GNSS, call details records, crowdsourcing, smartphone, sensing human activity, transportation planning

1. Introduction

The topic of smart cities gained increasing interest among researchers from different fields. The concept goes beyond the pure use of information and communication technologies (ICT) towards building smarter buildings, mobility solutions, sustainable living and smart governance that meets the needs of an urban population as a sustainable community. In this chapter, we examine the role and potential of sensing devices as one of key pillars towards smart mobility management. We particularly focus on the use of smartphones as ubiquitous sensing devices that provide more detailed insight into mobility behavior than ever before facilitating smarter mobility management, development of tailor made policy measures and advanced two-way communication channels between relevant stakeholders. To illustrate these potentials we report on the use of Routecoach app developed at Ghent University and used by more than 8000 users for mobility data collection in the city of Leuven, Belgium. We provide insights into lessons learned, ways in which collected data were used by different stakeholders, and identify future research needs that can alleviate existing gaps towards truly smart and seamless mobility management.

2. Sensing human activity

Understanding mobility behavior is one of the key elements in ensuring better transport and urban planning. Advances in these areas are welcomed as they can ensure more seamless mobility, which is particularly a demanding task in urban areas where different transport modes meet and often share same space. As mobility is service, and it is impossible to store its capacities at certain location for future time, when the service will be needed, but rather synchronized time-space respond to dynamic demand is needed. To be able to better estimate these demands, and provide adequate level of service, data on travel activities are collected. The traditional data collection process can be user-oriented or location-oriented.

2.1. User-oriented sensing

A user-oriented approach goes from starting point of mobility system's user and data collected this way are usually aggregated at the household level. This type of data collection process commonly involves implementation of paper or phone household surveys, or interviews, where people are asked to record or state their travel behavior on for instance an average weekday. Ideally, household travel surveys involve representative sample of target population, and processed data on trip origins and destinations, frequencies, purposes, and utilized transport modes serve as an input for transportation planning models. Ettema et al. [1] and Stopher and Greaves [2] have shown that data collected in this way deviated systematically from the actual travel behavior. Some examples of such deviation include tendency of the respondents to underreport non-motorized trips [3–5] or public transport users to overestimate their actual travel time [6]. Furthermore, response rates to these surveys tend to be low which represents challenge in terms of nonresponse bias [7]. To avoid these pitfalls paper travel diaries were introduced [8].

In paper travel diaries, one is asked to systematically note his or hers travel behavior details with respect to travel times, origin and destination locations, transport modes, trip purposes, and frequencies. The data collection interval is usually one complete week during non-holiday periods. Literature reports [9, 10] that respondents tend to postpone filling in these diaries, which results in obtaining incomplete and inconsistent information. Quite often this would include having trouble remembering and recording smaller trips (e.g., walking to nearby post office to pick up package delivery or to local library to return a book), rounding off time and distances [11], having difficulties in defining the exact locations of places they have visited, or underreporting multimodal trip segments (e.g., indicating trip made by public transport, but forgetting to mention walking to and from public transport stop or between metro and bus stops).

Indeed, both travel surveys and diaries were designed primarily to provide data for macroscopic traffic models, and respectively, were focused on capturing trips between traffic analysis zones (TAZs). A TAZ is the unit of geography most commonly used in conventional transportation planning models and represents spatially homogeneous land use area (e.g., residential area, industry area etc.). Size of the TAZ varies, but typically it is a zone of under 3000 people. Quite often, these zones match census block information which makes it easier to interpret models' outputs. As macroscopic traffic models are not focused on trips inside individual TAZs, but between different ones, both travel surveys and diaries ignored shorter trips within TAZs and, this way further impacted underreporting of smaller trips that were usually made by active transport modes [12, 13]. This resulted in bias in observed modal splits and further underpinned evolution of car-oriented transport.

2.2. Location-oriented sensing

Compared to user-oriented sensing, the location-oriented data collection process tries to capture travel entities that are passing predefined location. This can be one point in the transport network, but more often the data collection process includes several points dispersed geographically to cover target area and all input/output points to the target part of the network (e.g., main roads entering the city, train stations etc.). The most straightforward way is to manually note the number of vehicles or pedestrians that passed the predefined location within the predefined time interval (usually 15 min or 1 h). In addition, other traffic data like, vehicle occupancy rate or vehicle classifications can also be collected. As manual counting is quite expensive way to collect mobility data and suffers from human errors, this approach is further developed into automated counting of moving objects. For this purpose, different types of a data recorders and sensors placed on or under the traffic network surface can be used (e.g., pneumatic road tubes, piezoelectric sensors and inductive loops). This has been widely deployed over the past few decades but the implementation and maintenance costs tend to be high. In addition, they successfully extract only traffic counts while additional information stay unreported (e.g., vehicle occupancy rate). To avoid these pitfalls, video based techniques for traffic counting have been developed. They rely on vehicle identification and more advanced approaches can include automated vehicle classifications or capturing of vehicle occupancy rates.

2.2.1. Computer vision applications

For vehicle identification, usually license plate matching techniques are applied. These techniques consist of collecting vehicle license plate characters and arrival times at various checkpoints. Since manual collection of license plate information is less practical for high-speed roads, ideally this is done by video cameras and character recognition software to recognize and automatically transcribe the license plate number for subsequent computer processing

[14]. Collection of arrival times at different checkpoints makes it possible to process data in order to recreate vehicle movements (if data collection points are of adequate density) or to derive travel times in the transport network. However, the ability of video based method to correctly identify license plate characters is often influenced by factors such as vehicle speed, volume of vehicle flow, ambient illumination (day, night, sun, or shadow), spacing between vehicles (occlusion), weather conditions (rain, snow, fog), plate variety, physical position of the plate (tilt, rotation), etc. In general, the license plate capturing and recognition rates may vary from as low as 15% (for poor visibility/weather conditions) to as high as 85–90% [15]. Another application for mobility studies comes from the possibility of implementation of computer vision applications in vehicles to recognize the surroundings and adjust their driving behavior in line with this information. These types of applications are particularly interesting for automated vehicles as future mobility entities within the smart cities. **Figure 1** shows example of ongoing research activities under the Vebimobe project where applicability of computer vision for automated recognition of traffic signs within the city of Ghent, Belgium is studied. Specially designed vehicles test the ability to recognize traffic signs, while being integrated in traffic flow movements, from cameras that operate in different spectrums [16]. One of the main aims of the Vebimobe project is to examine readiness of related technologies in ensuring application of such data collection techniques for automated vehicles' speed adaption and more sustainable route guidance applications.

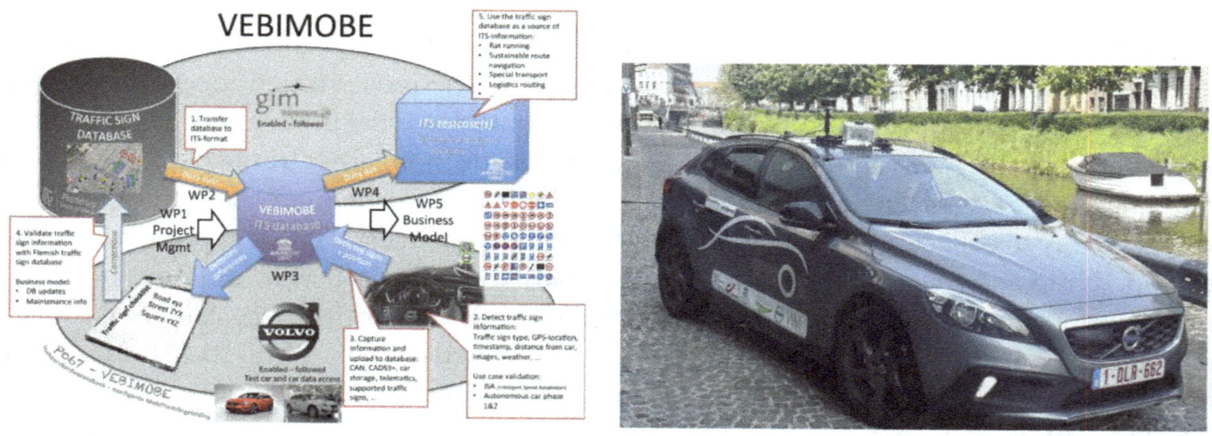

Figure 1. Vebimobe —organization structure of research activities (left) and test vehicle with equipment for computer vision supported detection of traffic signs (right) (from [16]).

Furthermore, for machines to be able to detect and identify people instead of vehicles (or traffic signs) is a more challenging task and sensing of humans has long been one of the hardest machine vision problems to tackle. Next to the inherited challenges with ambient illumination, occlusions (e.g. having umbrella), weather conditions (rain, snow, fog), etc., main challenge comes from wider diversity in appearance and more erratic way humans behave. So far successful applications mainly focus on recognizing human silhouettes and classifying activities (standing, walking) [17, 18] while face recognition performs better on a smaller scale. For urban areas where large number of people passes daily (e.g., in a train station), the performance is lower due to limits in recognition. However, for mobility-related applications where just separation between different transport modes is needed (in this case, just recognizing whether it is human/pedestrian), success rates are higher than in case where actually identifying a unique human is needed to, for example, compare travel times between successive locations.

2.2.2. Bluetooth scanning

More recently, Bluetooth has been suggested as an interesting alternative for location oriented sensing technology. Bluetooth is a wireless technology standard [19] for exchanging data over short distances. It uses short wavelength ultra-high frequency (UHF) radio waves in the industrial, scientific and medical (ISM) band from 2.4 to 2.485 GHz [20]. It was invented by telecom vendor Ericsson in 1994, and it can connect several devices, overcoming problems of synchronization which makes it particularly interesting for implementations ranging from fixed and mobile devices to building personal area networks [20]. Prior to the wireless connection of two devices through Bluetooth, the inquiry phase of the protocol needs to be completed. In this phase, an initiator device initiates the service discovery procedure by transmitting inquiry packets. Devices, that allow themselves to be discoverable, issue an inquiry response. The inquiry response includes information on device ID (48-bit identifier of the mobile device—MAC address) and clock [21]. The interesting feature, for mobility studies, is that these information are exchanged before any connection is established which allows completely unobtrusive sensing of nearby devices.

Today, Bluetooth has become an almost ubiquitous technology on modern mobile devices and private vehicle keys, by placing static Bluetooth sensors at strategic locations one can get insights into personal (based on mobile devices) or vehicle (based on keys) mobility in a variety of contexts. Due to the range limitations this technology is more appropriate for location-oriented tracking than user based one, but with additional processing user trajectories can be approximated based on the timestamp sequences. Phua et al. [22] have compared Bluetooth sensed data at supermarket and manually measured data using systematic sampling and found that trip lengths and user demographics were similar with the exception of underrepresenting older population. Other examples of sensing human mobility include travel time measurements of motorized traffic [23, 24], tracking of pedestrians [25], mobility-related incident detection [26, 27], dynamics at mass events [28] and others.

Figure 2 shows implementation of Bluetooth scanners for monitoring the crowd behavior during the Ghent Festivities in Ghent, Belgium (as described in [28]). Implementation aimed at supporting city event management for the organization, security, transport, and emergency service providers. Ghent Festivities take place, every year at the end of July, on 11 squares in the city center and lasts one full week (including both starting and ending weekends). Squares, and city itself, act as major attractions during this period hosting on-stage performances, food stands, and fairs that attract around two million people during festivities. On this occasion, 22 locations were covered with Bluetooth scanners. Collected data represented people's mobility within the festivity zone itself and the mobility to and from the festivity zone. Applications of the resulting data are manifold. The most direct result are the statistics about visitors and their sensed behavior (e.g., the number of visitors per day, the time and space distribution of visitors, the (sequence of) squares visited by individual visitors, etc.). A second, derived set of results deals with the distribution and dynamics of the crowd in the festivity zone and the city center. This information is vital for security services, which are monitoring the people density in order to plan safety measures as temporary closures of access to overcrowded squares or facilitating the circulation between certain festive locations. Derived information is also made available to visitors by the festivity app, assisting them to plan their visit avoiding overcrowded or temporary closed areas. A third set of results deals with the accessibility of the festivity zone and is derived from monitoring of the travel times between train stations, public transport stops, park and ride locations, and the festivity zone. For example, by analyzing sequence of Bluetooth scans, of the same IDs, starting from the park and ride facility towards the city center prolonged travel times can be observed. This suggests congestion problems on the route, where traffic police should intervene to facilitate the circulation of the public transport. This way, the sensed data assist partners to optimize safety and comfort to the visitors of the festivity [30].

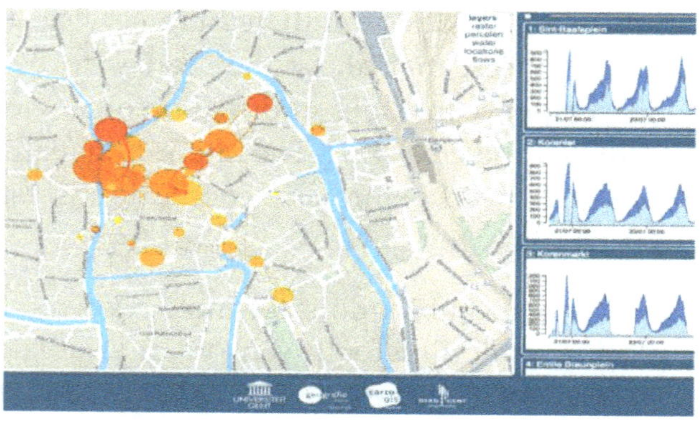

Figure 2. Bluetooth scanning implementation for mass events (Ghent Festivities event in Ghent, Belgium) [29].

The given example, illustrates the potential of using Bluetooth scanning for deriving origin and destination locations within the city or travel times. However, Bluetooth sensed data for mobility studies exhibit several limitations. First, sensed location is limited to the selected locations of static Bluetooth scanners. By analyzing sequence of observed devices' IDs between different locations, movement data can be estimated but exact paths are unknown unless Bluetooth scanners are placed at each intersection of the transport network. However, this might be quite expensive especially in large networks. Second issue is related to sample sizing and data quality. As only the activity of discoverable Bluetooth devices can be sensed, to report on the population level (e.g., absolute density or flow statistics) ratio of discoverable Bluetooth devices across general population needs to be determined. This is mainly done based on the manual counts of the total number of visitors at sensing locations; however, this process tends to be expensive.

Overall, when analyzing the implementation potential of location based sensing techniques for mobility studies, main limitations come from need for higher level of details, insights into utilized network connections and traffic flows dependencies, as well as need to include all users of the mobility network (pedestrians, bicyclists, public transport users, etc.). All of the location based sensing techniques score well for some of these challenges but fail at others. For example, sensors placed on or under the traffic network surface provide confident counts of vehicles, but cannot identify individual moving objects and therefore compare its observations across different locations. Computer vision based applications, have higher success rates in distinguishing between different transport modes, but still have limited success in identifying individual moving objects for practical implementation. Bluetooth sensors can easily identify individual devices and based on this information, track their moving sequences between different locations, but they require high density of sensing locations to reconstruct actual paths and cannot provide vehicle counts with same accuracy as road sensors or confident estimation of used transport modes.

2.3. Location-enabled devices

Introduction of location-enabled devices started an important revolution in mobility studies [31–34] as they allowed continuous tracking of movement locations and, this way, were able to fill some of the gaps that were present when collecting data using traditional methods [35–37]. Location-enabled devices mainly relay on global navigation satellite system (GNSS). The GNSS refers to a constellation of satellites providing signals from space transmitting positioning and timing data and, by definition, it provides global coverage. The GNSS allows small electronic receivers to determine their location (longitude, latitude, and altitude) to high precision. The signals also allow the electronic receiver to calculate the current local time to high precision, which allows time synchronization. Examples of GNSS include USA's NAVSTAR Global Positioning System (GPS) and Russia's Global'naya Navigatsionnaya Sputnikovaya Sistema (GLONASS) [38–40]. Europe is in the process of launching its own independent GNSS, Galileo, and China is currently expanding its regional BeiDou Navigation Satellite System into the global Compass navigation system [41]. First location-enabled devices that were used for mobility studies were usually installed in vehicles (**Figure 3**). Data collected in this way were used to note travel times [42, 43], detect congested segments in traffic network [44, 45], or reconstruct vehicle trajectories [46, 47]. Cai et al. [48] and Gullivera et al. [49] developed road traffic noise estimation models based on the collected GNSS data. Ćavar et al. [50, 51] used GNSS vehicle tracks to develop machine learning based model for predicting travel times in urban areas.

When collecting data on traffic stream for intelligent transportation system (ITS) applications, vehicles equipped with location-enabled devices are classified based on the vehicle driving styles as (a) average car (vehicle travels according to the driver's judgment of the average speed of the traffic stream); (b) floating car (driver "floats" with the traffic by attempting to safely pass as many vehicles as pass the test vehicle) and (c) maximum car (vehicle is driven at the posted speed limit unless impeded by actual traffic conditions or safety considerations). The information on the applied driving style is crucial for correct interpretation of the collected data and development of derived statistics. In the literature, the most often applied style is floating car [52, 53] as it has been seen to provide the most representative description of the actual traffic stream. However, since GNSS devices for mobility studies were usually installed in vehicles, consequently they only tracked a small portion of mobility behavior (i.e., car trips).

To track the full spectrum of mobility behavior, respondent needed to carry the handheld GNSS-devices continuously, as forgetting it would result in unreported gaps in the trip data [54]. This requires significant effort and discipline from the respondent. Furthermore, to be able to evaluate success rates of such data collection procedures, respondents often needed to note their trips manually which, together with carrying the device, represented significant burden to the respondents.

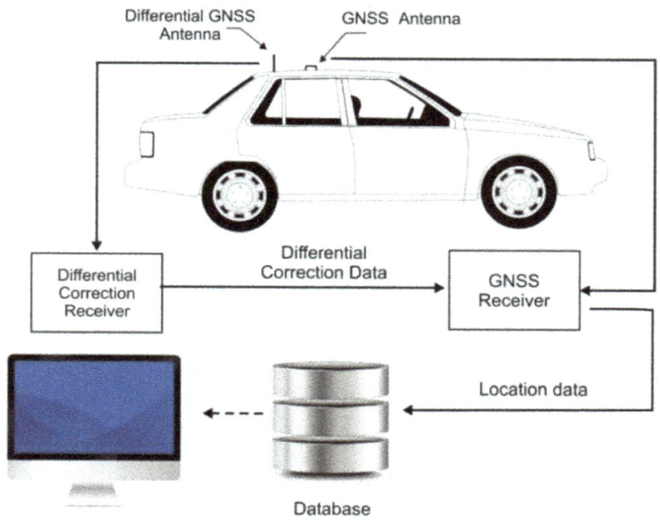

Figure 3. GNSS equipped vehicle (based on [15]).

2.4. Smartphone based crowdsourcing for mobility studies

Advances in development of GNSS chipsets allowed their integration in small devices, like smartphones, which resulted in emerging new possibilities for mobility behavior sensing. Carrying a smartphone has become a habit, and is therefore considered less of a burden, reducing the risk of non-reported trips. In addition, smartphones today have the same capabilities as the portable GNSS-device but also include additional sensors which can offer a more solid base for required interpretations of the data (e.g., use of accelerometer to determine the travel mode) and improve location precision.

In general, we can distinguish three ways in which mobile phone data are sensed for mobility studies (1) call detail record and network signalization data; (2) "passive" tracking and (3) "active" or "interactive" tracking.

2.4.1. Call detail record and network signalization data

Call detail record and network signalization data represents standardized data, collected by mobile network operators for billing purposes. Such data include records of all user-initiated activities such as calls, SMSs, internet, and data services where each record includes spatial and temporal parameters. In addition, network data include regular location updates of mobile devices, usually collected every hour, or every three hours, depending on the network generation and configuration. Handing over details (records created when user moves from area covered by one base station to another) are also noted. Therefore, frequency of the collected data varies depending on the device, network, and user activity. The location information is approximated to the telecom operator's base station points. The base stations (**Figure 4**) are land stations in the land mobile service that provide the connection between mobile phones and the wider telephone network. The size of area that each base station covers (and respectively, the distance between base stations) is not fixed and is a result of the trade-off between number of users (generated traffic), available frequencies, and quality of the service that operator wants to ensure. In practice, this results in a higher density of base stations in urban areas and lower in rural areas, but it is additionally influenced by build-up area and land configuration, as well as with specific user movement patterns in the vicinity of the base station (e.g., base stations that cover highways will have directed antennas, to ensure as little as possible handovers, and area that they cover will have highly elongated shape (**Figure 4**)). For these reasons, it is expected that location precision will be lower in rural areas and along high speed roads (**Figure 5**).

Figure 4. Base station (from [55]).

International telecommunication union reported 12-fold increase of penetration rates for mobile, and smartphone devices, since 2007 [56]. Such growth means that for the most of the areas (especially in developed countries) these data are capable to represent overall population movements. This potential gained much attention over the past years. Eurostat investigates possibility to replace some of the traditional data collection methods for general statistics with the use of call detail record and network signalization data [57–59]. Furthermore, their applicability in the scope of mobility studies has been investigated for rush hour analysis [60, 61], detection of variability in human activity spaces [62–64], correlation of mobility behavior with land use [65], and detection of TAZs and origin-destination pairs [66, 67]. Lui et al. [68] investigate possibility to develop validation measures for activity-based transportation models from mobile phone records. For this purpose, they approximated daily "home," "other," and "work"-related travel sequences and classified them to define activity-travel profiles. By comparing profiles with travel survey statistics, they demonstrated validation potential of the call detail records for this purpose. Gao and Liu [69] used the clustering technique to identify whether different phones travel in the same vehicle. They used mobile phone data to determine speed, vehicle counts, type, and density. This approach showed potential to be used for estimation of vehicle occupancies rates although manual counting would be needed to evaluate its effectiveness. Furthermore, Chen et al. [70] compared handover location updates and regular network based location updates to estimate travel speeds. AbdelAziz and Youssef [71] and Wang et al. [72] examined possibility to detect the transport mode one is using from their call detail record and base station location data.

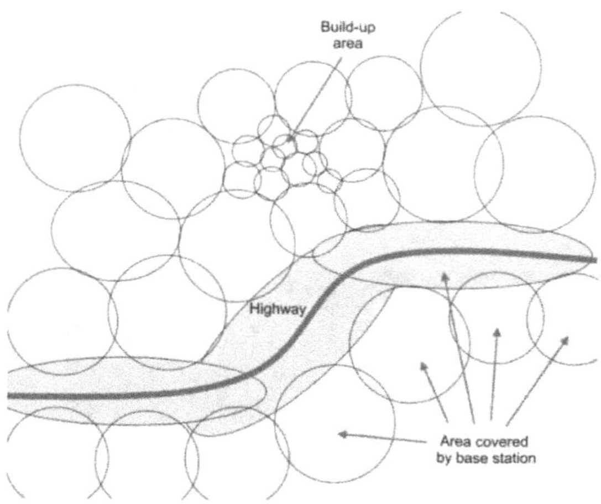

Figure 5. Base stations coverage.

At this point, all of the mobility-related studies that have examined possibility to use call detail record and network signalization data for analyzing the human travel behavior recognize high potential these data have for future applications. However, success rates achieved with processed data are still unsatisfactory for their practical implementation. Potentially, the largest limitation, in this sense, comes from low location precision (limited to cellular network base station locations) and time resolution (limited to users' activity or regular network location updates dependable upon the type/generation of the network). This makes these data more practical for extraction of origin and destination locations (which in this case would not overlap traditionally used TAZs but rather be based on the cellular network configuration) and crowd dynamics between different locations than for more detailed mobility studies. These solely are insufficient to replace traditional travel surveys but are a good starting point. Call detail record and network signalization data have a major advantage that comes from the fact that they are collected by all network operators, require no additional effort by users, no additional financial resources for their collection and cover wide areas, large populations and long time periods. On the other end, their usage for mobility, and other, studies at this point is hindered by a number of privacy and regulatory issues as well as some technological issues (e.g., how can the current data processing system be amended so that the processing of the mobile positioning data is also supported by statistical institutions), business related (e.g., operators see no benefits of providing data and, above all, are not motivated by possibility that concurrent companies have insights into their user base nor equipment locations), and methodological ones (e.g., the quality and applicability of the principles of statistical production in relation to mobile positioning data) [58, 73, 74].

2.4.2. "Passive" tracking

"Passive" tracking refers to the use of dedicated applications that run as a GNSS-based data logger in the background on the smartphone. Today, many applications collect such data (e.g., Google maps, Facebook, etc.). The use of "passive" tracked data is examined for the purposes of investigating individual mobility patterns [75, 76], speed analysis [77], traffic monitoring [78], or for large-scale sensing of human behavior for smart city-oriented applications [79]. Furthermore, smartphones are used as precise indoor positioning sensors in order to improve intelligent parking service [80] and as activity recognition sensors [81, 82]. Wan et al. [83] propose the use of mobile crowd sensing technology to support creation of dynamic route choices for drivers wishing to avoid congestion and Xia et al. [84] explore the use of smartphones, as sensors, for detection of transport modes from movement data of users.

The main advantage of this approach comes from higher spatial and temporal resolution of collected data than it is the case for mobile network call detail records. In fact, the spatial and temporal resolution of the collected data is set by the app maker itself, but can be influenced by user based on the mobile phone settings (e.g., positively by the use of GNSS, Wi-Fi and other network location data or negatively by simply turned-off mobile phone). Most often, the critical element in determining precision is the trade-off between phone's battery drain and data resolution. Foremski et al. [85] showed that smartphones can be used for crowd sensing with the decrease in battery lifetime by approximately 20%, which they found to be acceptable by users.

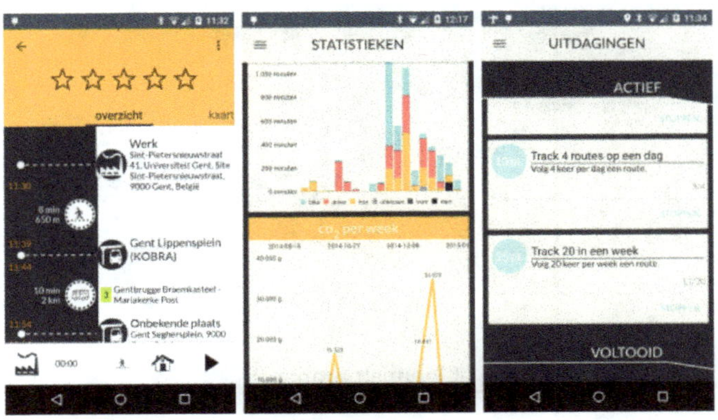

Figure 6. Routecoach — App.

Figure 6 shows Routecoach smartphone application [86, 87] that was developed at Ghent University [30] for collection of mobility data for the province of Flemish-Brabant in the frame of the Interreg IVb NWE project NISTO. The aim of NISTO (New Integrated Smart Transport Options) was to develop an evaluation and planning toolkit for mobility projects which is applicable transnationally and can be adopted by planners. The Leuven data collection process happened between January to April, 2015. In total, 8303 users actively participated by downloading the freely available application and collecting the data on one or more trips. Overall more than 30,000 trips have been recorded leading to about 350,000 km of recorded data (**Table 1**). The app had an option for "passive" data collection and "active" data collection (the "active" data collection segment of the app will be described in more details in the following section). **Figure 7** shows "passive" collected trips over the wider area of City of Leuven.

Variable	**Value**
Users	8303
Trips	30 000
Time period	4 months
GNSS points	3 960 234
km	340 000

Table 1. Sample descriptive data.

Figure 7. "Passively" logged trips (area: province of Flemish Brabant).

Applications of the resulting data are manifold. The most direct ones refer to user participation (e.g., general statistics) and mobility patterns (e.g., user activity). However, for detailed mobility studies significant post-processing is needed. This mainly refers to handling noisy data and removal of outliers. After data cleansing, map matching is required to match observed trips to the existing transport network locations. Care should be taken in this phase in order not to introduce errors by implemented map-matching algorithms and data quality control should be carried out with great care, as introduction of map-matching errors can lead to further errors in data interpretation and provide false base for mobility-related decision making.

Overall, main advantages of "passive" data collection, for mobility studies, compared to call detail record, come from higher spatial and temporal resolution. Compared to "active" tracking there is no need for interaction by respondents which reduces burden for the participant. That said, data collected this way require demanding data processing and interpretation efforts when compared to "active" tracking. Similar to call detail record processing advances, results of "passive" collected data processing are still not at mature level to replace travel diaries and surveys. One of the main challenges in this segment comes from the fact that it is hard to provide grand truth data, to "passive" logged data, and to check the success rates of the processing. As it is known that providing user with travel diary to note his trips will result in underreporting of small segments and trips made by active transport mode, these data are not applicable for representing the ground truth. In addition, the use of the apps is user initiated (user chooses to install, or not to install the app), whereas traditional data collection approaches were based on the initiative of the data collection institution. In this phase, data collection institution has an option to define representative sample and contact participants directly based on this definition. For mobile app data collection, it is challenging to determine the representatives of the sample as no background data are available about the user (e.g., no demographic data). It is always opted to aim for the law of large numbers, but aiming at mass data collection that would satisfy this condition would require substantial campaign resources and drastically increase the cost of data collection process. It is still to find the balance in this sense and tackle the question of crowdsourced data representativeness.

2.4.3. *"Active" and/or "interactive tracking"*

"Active" and/or "interactive tracking" represents the use of interactive mobile applications where respondents can report additional trip data as the start of the trip or transport mode. Such reporting was, for instance, used to investigate the influence of carbon dioxide emission information on mode choice [88] and, mostly, as ground truth for the development of supervised machine learning models in order to replace parts of traditional travel surveys [89, 90]. Semanjski and Gautama [91] examined applicability of "active" sensed mobility data to predict what transport mode one will use for the next trip (**Figure 8**). They applied gradient boosting trees and achieved a success rate of 73% indicating that such data can be used for smart city-oriented mobility services as provision of transport mode relevant pre-travel information or different incentives in order to impact one's mobility behavior towards more sustainable mode choices.

The use of "actively" logged data is also explored in inferring transport modes from mobile sensed data. These approaches strongly relay on GNSS records [35, 77], but also include data from other smartphone sensors [92, 93]. In many cases, these data are fused so that the GNSS data are used to improve accuracy of, for example, accelerometer-based approaches, or vice versa [32, 70, 84]. On average, literature reports successful recognition between three to five transport modes by using around four indicators [35, 94]. Recognized transport modes mainly include: motorized transport (without separation between personal vehicle and, e.g., bus), bike and walking, and their recognition relies on variables as speed and acceleration, implying that they give the highest indication of a transport mode [84, 92]. The main challenge arises from similar speeds obtained by more than one transport mode (e.g., bike and pedestrians, or private car and public transport) which is only partially solved at this point and additional knowledge is still needed to increase the accuracies (which is mainly below 90%). Overall, all studies tested the proposed approaches on limited time span of collected data (ranging from four hours to one week) and limited number of participants failing to capture wide range of longitudinal, e.g., monthly or yearly, variations in travel behavior patterns. In addition, such short time ranges imply observed behavior under similar conditions (e.g., weather condition) where potential limitations might lie in terms of transferability of developed approaches on a wider population and/or area.

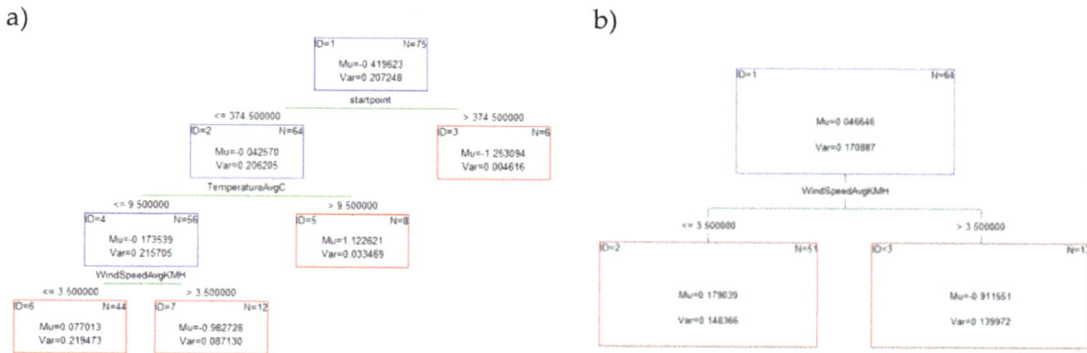

Figure 8. Decision trees for the transport modes (a) bike and (b) walk (from [89]).

For the Routecoach application, next to the "passive" logging that continuously tracked mobility behavior, participants were able to "actively" report and validate their data. "Active" data collection implied higher time-space resolution of the collected records and was initiated by the user. To start "active" data collection user needed to mark the transport mode used at the beginning of his or her trip. In addition, user was able to report the purpose of the trip, enabling extra contextual information. To reduce the burden to the participants, user-friendly graphical interface was developed so that users could simply switch between transport modes during their travels and, in this way, easily validate multimodal trips. To stop the "active" data collection user needed to mark end of the trip in the data collection app. In addition to the app, web interface was implemented (**Figure 9**) so that user can easily access personal mobility data (after the registration) and add or correct context of the trips (e.g., add purpose or correct wrongly introduced travel mode). In addition, web interface had incorporated web surveys that the user could fill in and provide personal information and insight into his or her attitudes toward different mobility options.

Data collected this way provide higher spatial and temporal resolution and rich (and validated) information on the context of travel activities. This significantly reduces need for data post-processing and allows relevant insights into mobility behavior. **Figure 10** shows Routecoach insights into observed delays at road network intersections in the city of Leuven, Belgium, providing local authorities with information on where to focus measures related to delay reductions. Insights on mobility behavior, at individual and aggregated levels, were also made available to the participants (personal data) and general audience (only aggregated results) so that everyone can adjust, if one wishes so, his or her behavior in order to avoid delays and crowded areas. High spatial and temporal resolution of data facilitated extraction of time relevant insights. Based on the crowdsensed data travel time for different transport modes could be observed and impact of newly introduced measures evaluated. For example, **Figure 11** shows bike travel time isochrones, where impact of new bike highway can be easily noticed in the North, and then North-East part of the network (as bike highway changes its direction). In addition, comparison of different transport modes is enabled as their performance can be simultaneously confronted. **Figure 12** shows accessibility of the main train station in Leuven during the afternoon peak hour. Blue area marks parts of the city from which it is faster to reach train station during this period than by car. Red areas indicate regions from which one would reach train station faster by car. These insights engaged citizens and policy maker into constructive discussion on mobility options and enable smarter mobility management.

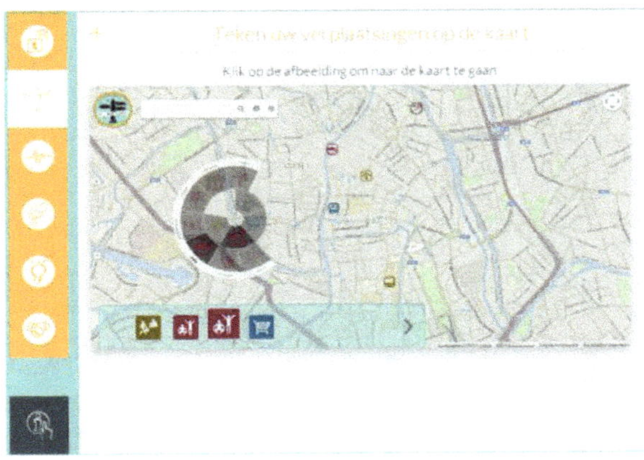

Figure 9. Routecoach – web interface.

Figure 10. Delays at transport network intersections.

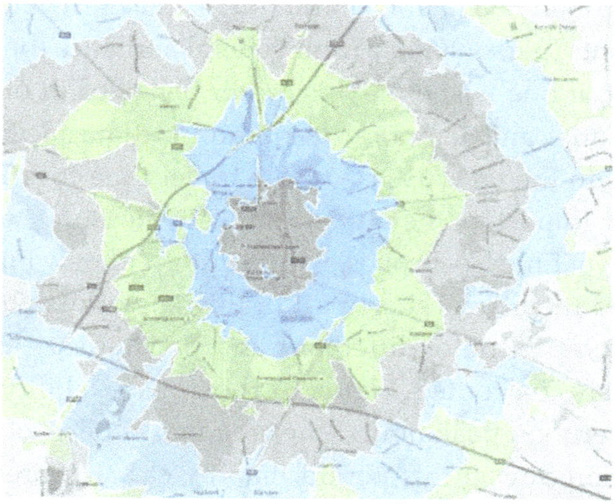

Figure 11. Bike travel time isochrones.

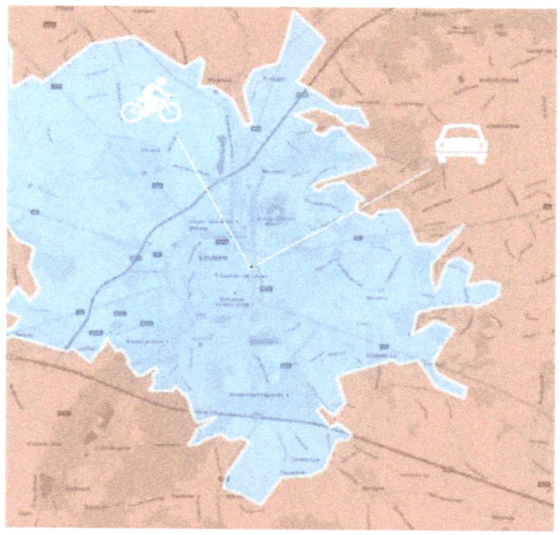

Figure 12. Accessibility of the main train station during afternoon peak hour.

Although "active" logging requires manual intervention by the respondent, this burden seems to be limited because the reporting is restricted to short entries at the very moment of departure and arrival. As a consequence, time and location of the departure and arrival can be more accurately detected, and there is no need for demanding data processing as splitting GNSS-based track into parts travelled by different modes [35, 95]. Overall, "active" data logging overcomes some of the weaknesses of call detail record and "passive" data collection approaches. For one, it provides trip context and reduces the need for extensive data post-processing. In addition, it also offers ground truth data for development of different machine learning based algorithms that can evolve towards the transport mode, or trip purpose, recognized from "passive" logged data. This way, more seamless transition from traditional data collection approaches, as travel surveys and diaries, towards fully data driven mobility management is facilitated. Another advantage comes from user validated data, and its potential to find balance between campaign expenses (to familiarize users with the data collection and app itself) and need for the representative sample, as based on the user provided personal information, one can extract representative subsample from the overall dataset. This can significantly reduce the cost of mobility data collection and creation of verified inputs for transport planning models. Compared to call details record, main advantages of "active" logged data come from higher spatial and temporal resolution. An example of this can be seen in quite demanding task to join data of lower resolution with, for example, freely available data on land use. Land use data have been often implemented to estimate trip purpose. Therefore determining whether trip ended at the school or office location is a quite challenging task, based on the call detail records, as within the area covered by one base station potentially there are both education, residential, work and commercial facilities. On the other end, the main challenge for "active" data collection comes from user engagement, trip reporting discipline, and motivation to participate in such activities. Although, users provide validated data on volunteering bases on same details as they were asked in traditional travel diaries, if existing, their privacy-related concerns need to be addressed. Transparent data processing and usage, as well as evident benefits in terms of better mobility management seem to be strong advocates for user motivation and participation.

3. Conclusion

The introduction of smartphones as mobility sensing devices exhibits multiple advantages when compared to traditional data collection approaches. It reduces the number of unreported trips which was the case for travel diaries and surveys where users often postponed completing these to later on during the day or week. This resulted in making it hard to remember short trips (e.g., walk to nearby restaurant during the lunch break). Regarding the mobility management, the above mentioned reflected as underrepresentation of walking and biking trips providing false insights into existing modal splits and supporting favoritism towards car-oriented transportation planning. In this sense, the use of smartphones can support more balanced sensing of mobility behavior across the use of different transport modes. In addition, as carrying a smartphone has become a habit for many people, the issue of unreported gaps in the trip data is overcome. Nevertheless the use of "active" logging for smart city-oriented mobility applications is advised as knowledge discovery from "passive" logged data remains unsatisfying (e.g., real time splitting of trips at transport mode changing points or estimation of trip purposes from "passively" collected data). This brings forward challenges related to respondents' motivation and participation in "active" logging. In this regard, the use of different incentives is still being researched [96]. So far, adjustable and personalized rewarding systems, social networks based interaction and gamifications show the highest potential. But, this area still remains to be further explored in order to relate these with different user profiles and balance between incentives and personal motivation. Regarding different user profiles, their role is of the most value when considering smartphones as tools for policy makers to deliver personalized mobility-related messages and make targeted policy measures. Psychological studies in this field suggest that profiling respondents based on their attitudes towards sustainable mobility options shows good potential in initiating behavioral change. In this context, smartphones can be used both as sensing devices and as two-way communication tools where targeted, time-space, relevant information can be delivered to users (e.g., reported estimated delays on the foreseen route of interest). This way, users can make more informed mobility decisions and information on observed behaviors can be integrated into advanced mobility management systems.

Author details

Ivana Semanjski* and Sidharta Gautama

*Address all correspondence to: isemanjs@ugent.be

Ghent University, Department of Telecommunications and Information Processing, Ghent, Belgium

References

[1] Ettema, D., Timmermans, H. & Van Veghel, L., 1996. Effects of Data Collection Methods in Travel and Activity Research. Eindhoven, Netherlands: European Institute of Retailing and Services Studies.

[2] Stopher, P. & Greaves, S., 2007. Household travel surveys: where are we going? Transportation Research Part A: Policy and Practice, 41(4), 367–381.

[3] Forsyth, A., Krizek, K. J. & Weinstein Agrawal, A., 2010. Measuring Walking and Cycling Using the PABS (Pedestrian and Bicycling Survey) Approach: A Low-Cost Survey Method for Local Communities, San Jose: Mineta Transportation Institute.

[4] Itoh, S. & Hato, E., 2013. Combined estimation of activity generation models incorporating unobserved small trips using probe person data. Journal of the Eastern Asia Society for Transportation Studies, 10, 525–537.

[5] Weinstein, A. & Schimek, P., 2005. How Much Do Americans Walk? An Analysis of the 2001 NHTS. s.l., Transportation Research Board.

[6] Clifton, K. & Muhs, C., 2012. Capturing and representing multimodal trips in travel surveys. Transportation Research Record: Journal of the Transportation Research Board, 2285, 74–83.

[7] Sharp, J. & Murakami, E., 2001. Travel Survey Methods and Technologies Resource Paper. Washington: United States Census Report.

[8] Stopher, P. & Wilmot, C., 2002. Some new approaches to designing household travel surveys–time-use diaries and GPS. Washington, Paper Presented at the 79th Annual Meeting of the Transportation Research Board.

[9] Arentze, T. et al., 2001. New activity diary format: design and limited empirical evidence. Transportation Research Record: Journal of the Transportation Research Board, 1768(1), 79–88.

[10] Groves, R., 2006. Nonresponse rates and nonresponse bias in household surveys. Public Opinion Quarterly, 70(5), 646–675.

[11] Witlox, F., 2007. Evaluating the reliability of reported distance data in urban travel behaviour analysis. Journal of Transport Geography, 15(3), 172–183.

[12] Declercq, K., Janssens, D. & Wets, G., 2013. Travel survey. Diepenbeek: Transportation Research Institute, Universiteit Hasselt.

[13] Saelens, B., Sallis, J. & Frank, L., 2003. Environmental correlates of walking and cycling: Findings from the transportation, urban design, and planning literatures. Annals of Behavioral Medicine, 25(2), 80–91.

[14] Jiao, J., Ye, Q. & Huang, Q., 2009. A configurable method for multi-style license plate recognition. Pattern Recognition, 42(3), 358–369.

[15] Ćavar, I., 2010. Estimation of travel time in urban areas based on fused spatio-temporal and meteorological data. Zagreb: University of Zagreb.

[16] De Mol, J. et al., 2016. Vebimobe: accurate speed information for proper driving. Verkeersspecialist, 227, 20–23.

[17] Bredereck, M., Jiang, X., Körner, M. & Denzler, J., 2012. Data association for multi-object tracking-by detection in multi-camera networks. s.l., Proceedings of the Sixth International of Conference.

[18] Morbée, M., Tessens, L., Aghajan, H. & Philips, W., 2010. Dempster-Shafer based multi-view occupancy maps. Electronics letters, 46(5), pp. 341–342.

[19] Institute of Electrical and Electronics Engineers, 2002. 802.15.1-2002—IEEE Standard for Telecommunications and Information Exchange Between Systems. New York: IEEE.

[20] Bluetooth SIG, 2016. Basic rate/enhanced data rate (br/edr). [Online] Available at: https://www.bluetooth.com/what-is-bluetooth-technology/bluetooth-technology-basics/br-edr [Accessed 22 June 2016].

[21] Bourk, T., Howes, T. & Seymour, B., 2008. Discovery Whitepaper. Kirkland: Bluetooth SIG.

[22] Phua, P., Page, B. & Bogomolova, S., 2015. Validating Bluetooth logging as metric for shopper behaviour studies. Journal of Retailing and Consumer Services, 22, 158–163.

[23] Park, H. & Haghani, A., 2015. Optimal number and location of Bluetooth sensors considering stochastic travel time prediction. Transportation Research Part C: Emerging Technologies, 55, 203–216.

[24] Portugais, B. & Khanal, M., 2014. Adaptive traffic speed estimation. Procedia Computer Science, 32, 356–363.

[25] Utsch, P. & Liebig, T., 2012. Monitoring Microscopic Pedestrian Mobility Using Bluetooth. Guanajuato, Intelligent Environments (IE), 2012 8th International Conference on, IEEE.

[26] Bullock, D. M., Haseman, R., Wasson, J. S. & Spitler, R., 2010. Anonymous Bluetooth Probes for Measuring Airport Security Screening Passage Time: The Indianapolis Pilot De-ployment. Washington: Transportation Research Board.

[27] Margreiter, M., 2016. Automatic incident detection based on bluetooth detection in Northern Bavaria. Transportation Research Procedia, 15, 525–536.

[28] Versichele, M., Neutens, T., Delafontaine, M. & Van de Weghe, N., 2012. The use of Bluetooth for analysing spatiotemporal dynamics of human movement at mass events: A case study of the Ghent Festivities. Applied Geography, 32(2), 208–220.

[29] Ghent University, 2016. MOVE. [Online] Available at: http://move2.ugent.be/ [Accessed 20 March 2016].

[30] Gautama, S. et al., 2017 (In press). Observing Human Activity Through Sensing. In: V. Loreto, et al. eds. Participatory Sensing, Opinions and Collective Awareness. s.l. Switzerland: Springer International Publishing, pp. 47–68

[31] Ćavar, I., Marković, H. & Gold, H., 2006. GPS Vehicles Tracks Data Cleansing Methodology. s.l. International Conference on Traffic Science.

[32] Feng, T. & Timmermans, H., 2014. Extracting activity-travel diaries from GPS data: Towards integrated semi-automatic imputation. Procedia Environmental Sciences, 22, 178–185.

[33] Wolf, J., Bricka, S., Ashby, T. & Gorugantua, C., 2004. Advances in the Application of GPS to Household Travel Surveys. Washington: National Household Travel Survey Conference.

[34] Wolf, J., Guensler, R. & Bachman, W., 2001. Elimination of the travel diary: Experiment to derive trip purpose from global positioning system travel data. Transportation Research Record: Journal of the Transportation Research Board, 1768, 125–134.

[35] Bohte, W. & Kees, M., 2009. Deriving and validating trip purposes and travel modes for multi-day GPS-based travel surveys: A large-scale application in the Netherlands. Transportation Research Part C: Emerging Technologies, 17, 285–297.

[36] Bricka, S., Sen, S., Paleti, R. & Bhat, C., 2012. An analysis of the factors influencing differences in survey-reported and GPS-recorded trips. Transportation Research Part C: Emerging Technologies, 21(1), 67–88.

[37] Montini, L. et al., 2015. Comparison of travel diaries generated from smartphone data and dedicated GPS devices. Transportation Research Procedia, 11, 227–241.

[38] Misra, P. & Enge, P., 2006. Global Positioning System: Signals, Measurements and Performance. Second Edition ed. Lincoln: Ganga-Jamuna Press.

[39] Hofmann-Wellenhof, B., Lichtenegger, H. & Collins, J., 2012. Global Positioning System: Theory and Practice. s.l. Springer Science & Business Media, Wien, Austria.

[40] Zinoviev, A. E., 2005. Using GLONASS in combined GNSS receivers: current status. s.l. Proceedings of ION GNSS.

[41] Groves, P. D., 2013. Principles of GNSS, Inertial, and Multisensor Integrated Navigation Systems. s.l. Artech House, Norwood, USA.

[42] Gallagher, J., 1996. Travel Time Data Collection using GPS. Albuquerque, National Traffic Data Acquisition Conference, The Federal Highway Administration, Washington, USA.

[43] Roden, D. B., 1996. GPS-Based Travel Data Collection. Kansas City, Proceedings of the 9th International Technical Meeting of the Satellite Division of The Institute of Navigation, The Institute of Navigation, Manassas, USA.

[44] Wen, H., Sun, J. & Zhang, X., 2014. Study on Traffic Congestion Patterns of Large City in China Taking Beijing as an Example. Shaoxing, The 9th International Conference on Traffic and Transportation Studies, Elsevier, Amsterdam, Netherlands.

[45] Xiaolin Panga, L., Chawla, S., Liu, W. & Zheng, Y., 2013. On detection of emerging anomalous traffic patterns using GPS data. Data & Knowledge Engineering, 87, 357–373.

[46] Jose, D., Prasad, S. & Sridhar, V., 2015. Intelligent vehicle monitoring using Global Positioning System and Cloud Computing. Procedia Computer Science, 50, 440–446.

[47] Kim, J.-H. & Oh, J.-H., 2000. A land vehicle tracking algorithm using stand-alone GPS. Control Engineering Practice, 8(10), 1189–1196.

[48] Cai, M., Zou, J., Xie, J. & Ma, X., 2015. Road traffic noise mapping in Guangzhou using GIS and GPS. Applied Acoustics, 87, 94–102.

[49] Gullivera, J. et al., 2015. Development of an open-source road traffic noise model for exposure assessment. Environmental Modelling & Software, 74, 183–193.

[50] Ćavar, I., Kavran, Z. & Bosnjak, R., 2013. Estimation of travel times on signalized arterials. Journal of Civil Engineering and Architecture, 7(9), 1141–1149.

[51] Ćavar, I., Kavran, Z. & Rapajic, R. M., 2012. Travel Time Estimation Results with Supervised Non-parametric Machine Learning Algorithms. Barcelona, The First International Conference on Data Analytics, International Academy Research and Industry Association, Wilmington, USA.

[52] Chen, F., Shen, M. & Tang, Y., 2011. Local path searching based map matching algorithm for floating car data. Procedia Environmental Sciences, 10(A), 576–582.

[53] Yong-Chuan, Z., Xiao-Qing, Z., li-ting, Z. & Zhen-ting, C., 2011. Traffic congestion detection based on GPS floating-car data. Procedia Engineering, 15, 5541–5546.

[54] Turner, S., Eisele, W., Benz, R. & Holdener, D., 1998. Travel Time Data Collection Handbook. Arlington: Texas Transportation Institute.

[55] New era energy, 2016. New era energy. [Online] Available at: http://www.neenigeria.com/ [Accessed 08 June 2016].

[56] International telecomunication union, 2015. ICT Facts and Figures 2015, Geneva: ITU.

[57] Eurostat, 2013. Feasibility Study on the Use of Mobile Positioning Data for Tourism Statistics, Luxemburg: Eurostat.

[58] Eurostat, 2013. Feasibility Study on the Use of Mobile Positioning Data for Tourism Statistics, Task 1. Stock-taking, Luxemburg: Eurostat.

[59] Eurostat, 2014. Feasibility Study on the Use of Mobile Positioning Data for Tourism Statistics Consolidated Report, Luxembourg: Eurostat.

[60] Bar-Gera, H., 2007. Evaluation of a cellular phone-based system for measurements of traffic speeds and travel times: A case study from Israel. Transportation Research Part C: Emerging Technologies, 15(6), 380–391.

[61] Järv, O. et al., 2012. Mobile phones in a traffic flow: A geographical perspective to evening rush hour traffic analysis using call detail records. PLoS One, 7(11), e49171.

[62] Gonzalez, M., Hidalgo, C. & Barabasi, A.-L., 2008. Understanding individual human mobility patterns. Nature, 453, 779–782.

[63] Hoteit, S., Secci, S. & Sobol, S., 2014. Estimating human trajectories and hotspots through mobile phone data. Computer Networks, 64, 296–307.

[64] Järv, O., Ahas, R. & Witlox, F., 2014. Understanding monthly variability in human activity spaces: A twelve-month study using mobile phone call detail records. Transportation Research Part C: Emerging Technologies, 83, 122–135.

[65] Toole, J., Ulm, M. & González, M., 2012. Inferring land use from mobile phone activity. Beijing, Proceedings of the ACM SIGKDD International Workshop on Urban Computing, ACM New York, New York, USA.

[66] Alexander, L., Jiang, S., Murga, M. & González, M., 2015. Origin–destination trips by purpose and time of day inferred from mobile phone data. Transportation Research Part C: Emerging Technologies, 58, 240–250.

[67] Trasarti, R. et al., 2015. Discovering urban and country dynamics from mobile phone data with spatial correlation patterns. Telecommunications Policy, 39(3-4), 347–362.

[68] Liu, F. et al., 2014. Building a validation measure for activity-based transportation models based on mobile phone data. Expert Systems with Applications, 41(14), 6174–6189.

[69] Gao, H. & Liu, F., 2013. Estimating freeway traffic measures from mobile phone location data. European Journal of Operational Research, 229(1), 252–260.

[70] Chen, C.-H. et al., 2013. Traffic speed estimation based on normal location updates and call arrivals from cellular networks. Simulation Modelling Practice and Theory, 35, 26–33.

[71] AbdelAziz, A. M. & Youssef, M., 2015. The Diversity and Scale Matter: Ubiquitous Transportation Mode Detection using Single Cell Tower Information. 2015 IEEE 81st Vehicular Technology Conference (VTC Spring), 1–5.

[72] Wang, H., Calabrese, F. & Di Lorenz, G., 2010. Transportation Mode Inference from Anonymized and Aggregated Mobile Phone Call Detail Records. Madeira, 13th International IEEE Annual Conference on Intelligent Transportation Systems, IEEE, New York, USA.

[73] Seidl, D., Jankowski, P. & Ming-Hsian, T., 2015. Privacy and spatial pattern preservation in masked GPS trajectory data. International Journal of Geographical Information Science, 30(4), 785–800.

[74] Vij, A. & Shankari, K., 2015. When is big data big enough? Implications of using GPS-based surveys for travel demand analysis. Transportation Research Part C: Emerging Technologies, 56, 446–462.

[75] Calabrese, F., Diao, M. & Di Lorenzo, G., 2013. Understanding individual mobility patterns from urban sensing data: A mobile phone trace example. Transportation Research Part C: Emerging Technologies, 26, 301–313.

[76] Liu, J. et al., 2012. iParking: An Intelligent Indoor Location-Based Smartphone Parking Service. Sensors, 12(11), 14612–14629.

[77] Huss, A., Beekhuizen, J., Kromhout, H. & Vermeulen, R., 2013. Using GPS-derived speed patterns for recognition of transport modes in adults. International Journal of Health Geographics, 13(49), pp. 1–8.

[78] Herrera, J. et al., 2010. Evaluation of traffic data obtained via GPS-enabled mobile phones: The Mobile Century field experiment. Transportation Research Part C: Emerging Technologies, 18(4), 568–583.

[79] Macias, E., Suarez, A. & Lloret, J., 2013. Mobile sensing systems. Sensors, 13(12), 17292–17321.

[80] Liu, F., Janssens, D., Wets, G. & Cools, M., 2013. Annotating mobile phone location data with activity purposes using machine learning algorithms. Expert Systems with Applications, 40, 3299–3311.

[81] Parviainen, J. et al., 2014. Adaptive activity and environment recognition for mobile phones. Sensors, 14(11), 20753–20778.

[82] Shoaib, M. et al., 2015. A survey of online activity recognition using mobile phones. Sensors, 15(1), 2059–2085.

[83] Wan, J. et al., 2016. Mobile crowd sensing for traffic prediction in internet of vehicles. Sensors, 16(1), 1–15.

[84] Xia, H., Qiao, Y., Jian, J. & Chang, Y., 2014. Using smart phone sensors to detect transportation modes. Sensors, 14(11), 20843–20865.

[85] Foremski, P., Gorawski, M., Grochla, K. & Polys, K., 2015. Energy-efficient crowdsensing of human mobility and signal levels in cellular networks. Sensors, 15(9), 22060–22088.

[86] Google Api, 2015. Activity recognition Api. [Online] Available at: https://developers.google.com/android/reference/com/google/android/gms/location/ActivityRecognition [Accessed 12 December 2014].

[87] Routecoach, 2015. Routecoach. [Online] Available at: http://www.routecoach.be/ [Accessed 20 June 2015].

[88] Brazil, W. & Caulfield, B., 2013. Does green make a difference: The potential role of smartphone technology in transport behaviour. Transportation Research Part C: Emerging Technologies, 37, 93–101.

[89] Nitsche, P., Widhalm, P. & Breuss, S., 2012. A strategy on how to utilize smartphones for automatically reconstructing trips in travel survey. Procedia—Social and Behavioral Sciences, 48, 1033–1046.

[90] Nitsche, P. et al., 2014. Supporting large-scale travel surveys with smartphones—A practical approach. Transportation Research Part C: Emerging Technologies, 43, 212–221.

[91] Semanjski, I. & Gautama, S., 2015. Smart city mobility application—Gradient boosting trees for mobility prediction and analysis based on crowdsourced data. Sensors, 15, 15974–15987.

[92] Hemminki, S., Nurmi, P. & Tarkoma, S., 2013. Accelerometer-Based Transportation Mode Detection on Smartphones. Rome, Proceedings of the 11th ACM Conference on Embedded Networked Sensor Systems, ACM New York, New York, USA.

[93] Manzoni, V., Maniloff, D. & Kloeck, K., 2010. Transportation Mode Identification and Real-time CO2 Emission Estimation Using Smartphones. Cambridge: SENSEable City Lab, Massachusetts Institute of Technology.

[94] Reddy, S. et al., 2008. Determining transportation mode on mobile phones. Washington, 12th IEEE International Symposium on Wearable Computers, IEEE, New York, USA.

[95] Safi, H., Assemi, B., Mesbah, M. & Ferreira, L., 2016. Trip detection with smartphone-assisted collection of travel data. Transportation Research Record: Journal of the Transportation Research Board, 2594, 18–26.

[96] Semanjski, I., Lopez Aguirre, A.J., De Mol, J., Gautama, S. 2016. Policy 2.0 platform for mobile sensing and incentivized targeted shifts in mobility behaviour. Sensors, 16.

7

Photonics for Smart Cities

Joseph S.T. Smalley, Felipe Vallini,

Abdelkrim El Amili and Yeshaiahu Fainman

Abstract

We review the current applications of photonic technologies to Smart Cities. Inspired by the future needs of Smart Cities, we then propose potential applications of advanced photonic technologies. We find that photonics already has a major impact on Smart Cities, in terms of smart lighting, sensing, and communication technologies. We further find that advanced photonic technologies could lead to vastly improved infrastructure, such as smart water-supply systems. We conclude by proposing directions for future research that will have the greatest impact on realizing Smart City initiatives.

Keywords: photonics, nanophotonics, nanotechnology, Smart Cities, sensors, pollution, water supply, infrastructure, metamaterials, nanolasers, optics

1. Introduction

Cities behave as complex and adaptive systems that both require and inspire technology [1, 2]. Photonics—the scientific and engineering discipline dedicated to the generation, transmission, processing, and detection of light—enables much of the information and communication technology that make cities smarter. Nanoscale photonics, also known as nanophotonics, in particular, delivers advanced technologies for improving the quality of life of city inhabitants. In this chapter, we ask and answer two main questions: (1) How are current photonic technologies contributing to the development of Smart Cities? and (2) how can the Smart Cities paradigm inspire a new generation of photonic technologies?

We have surveyed the existing literature on both Smart City initiatives and applications of photonic technologies, with the aim of integrating our findings into a coherent perspective of the current and potential impact of photonics on Smart Cities.

The chapter is divided as follows. In Section 2, we present our conceptual framework, which identifies relationships between photonics, Smart Cities, and complexity science [3]. We also provide a brief overview of the many applications of photonics in the context of urban development. In Section 3, we address our first primary question in detail. In order to achieve sufficient depth, we focus on several existing application areas of photonics. Namely we focus on smart lighting for human-centric illumination and urban agriculture; smart sensor arrays for environmental and resource consumption monitoring; and smart optical communication and signal processing systems. In Section 4, we address our second primary question in detail, using recent developments in urban water management, that is, the Flint water crisis and Southern California drought, as real-world examples. Again, to achieve sufficient depth, we focus on an exemplary potential next-generation photonic technology, a smart water sensing network. Finally, we propose avenues for further photonics research inspired by the needs of Smart Cities. It should be emphasized that it is impossible to cover all current and potential applications of photonics to Smart City technologies. We do our best to focus on what we believe are, either the most vital to improving urban quality of life, or the least well known to the research community.

2. Conceptual framework

Firstly, we propose a conceptual framework that will guide us throughout this chapter, and beyond, which is illustrated in **Figure 1**. Photonics provides technologies that enable the growth of social networks, the internet of things (IOT) [4], and maintenance of infrastructure, among other applications. Conversely, Smart Cities provide applications for photonics and drive advancement of future generations of materials, devices, circuits, and systems. Simultaneously, data collected by ubiquitous sensor arrays in Smart Cities may be delivered and analyzed not just for immediate actuation but also to researchers who study and predict phenomena that need to be monitored or controlled. Scientists may analyze the data to understand cities as complex adaptive systems of systems (CASoS), which are social–technical–natural networks exhibiting highly nonlinear dynamics [1]. Problems in CASoS are often formalized as optimization problems that require large amounts of processing power. While digital electronic number-crunching is today's norm, future processing of certain problems may be better served by optical and optoelectronic accelerators and/or signal processing systems, in which coupled photonic elements model complex dynamical behavior and augment the capabilities of electronic processors [5]. Solving these problems provides understanding to cities. Additionally, complexity science may inform photonics by solving many-body problems at the atomic, nanoscales, and mesoscales [3]. In this chapter, we focus on the interaction between photonics and Smart Cities. The other interactions illustrated by **Figure 1** form part of our roadmap for future research (Section 4).

While the definition of a Smart City remains somewhat ambiguous, researchers and practitioners do seem to be converging on a common idea [6]. Common features of the various definitions include emphasis on management and organization, technology, governance, policy, people, economy, built infrastructure, and the natural environment. Herein, we focus on technology and loosely follow the definition of Harrison *et al.*, who described the Smart City as being instrumented, integrated, and intelligent [7]. In more concrete terms, all Smart Cities must have certain components, namely data collection, integration, and analysis, all leading to some form of decision making that actuates the sensed environment (**Figure 2**). We follow the description of Dirk *et al.*, whereby we view the sensed environment as a system of systems, individual systems being people, businesses, communications, transport, water, and energy [8]. Additionally, we add the atmosphere to this list. The entire system is a CASoS [1]. While the size of cities and outward appearance may be extremely different, the properties of cities appear to exhibit common scaling properties that may be understood using the tools of complexity theory [9, 10]. Additionally, recommendations from mayors and researchers participating in IBM's Smarter Cities challenge have common prevailing themes, despite apparent differences in city size, location, and history [11, 12].

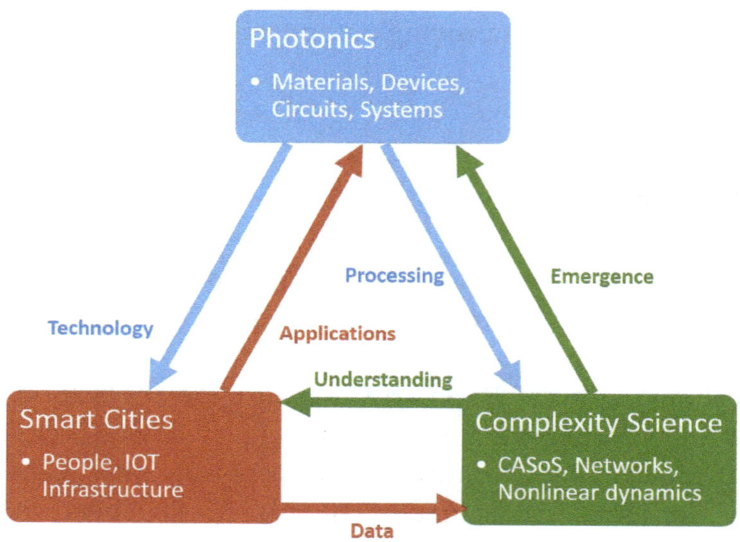

Figure 1. Conceptual framework relating photonics, Smart Cities, and complexity science.

In the context of Smart Cities, photonic sensors and phased arrays enabled data collection from the environment, while communications technologies enable high bandwidth connectivity between all Smart City components. As evidenced by the recently formed American Institute of Manufacturing for Integrated Photonics [13], key application areas of integrated photonics include (1) digital communications within datacenters and between data centers and end-users; (2) analog radio frequency (RF) and microwave communication with fiber optic links; (3) chip-scale chemical and biochemical sensing; and (4) light detection and ranging (LIDAR). Generally, we may classify the first two key applications as communications and signal processing and the last two as sensing modalities (**Figure 2**). Ultimately, through analysis and visualization, decisions are made by both humans and machines that act upon the environment, creating a feedback loop for sensing and actuation of the environment. Additionally, photonics provides lighting for cities, which affects all systems, enabling operation at all hours, and increasingly completely new possibilities, such as energy-efficient vertical farming.

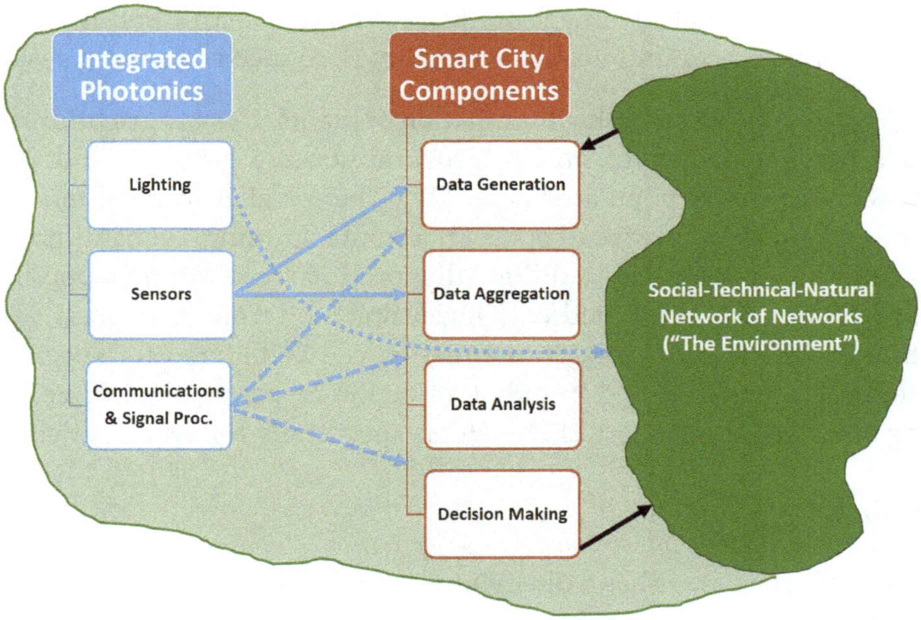

Figure 2. Schematic illustrating general relationships between major applications of integrated photonics and components of Smart Cities.

3. Current applications of photonics to Smart Cities

We broadly classify current major applications of photonics to Smart Cities as (1) lighting, (2) sensors, and (3) communications and signal processing. In the following, we briefly review examples of these application areas.

3.1. Smart lighting

The replacement of incandescent urban lighting for illumination and traffic control with light-emitting diodes (LEDs) is an ongoing application of photonics [14, 15]. LEDs are significantly more energy efficient than older light sources, and their emission wavelength is easily tuned from the UV to infrared through appropriate choice of semiconducting compounds. Several key applications of LEDs that we discuss briefly are human-centric lighting [16] and vertical farming [17].

While often taken for granted, the lighting conditions around residents of a city affect their mood, productivity, sleep patterns, and visual acuity [16]. Human-centric lighting has emerged to address the growing body of knowledge on the effects of lighting on human behavior. Case studies have shown improved learning of school children [18] and increased patient satisfaction in hospitals, when human-centric lighting replaced conventional, static indoor lighting conditions. Human-centric lighting requires dimmable light sources of different emission colors and was originally implemented with fluorescent tubes. However, tunable and dimmable LEDs with an adaptive spectrum and superior performance have become the preferable light source [16]. An additional advantage of LEDs for human-centric lighting is their inherent potential with sensing and communication capability in connected lighting systems [19].

Smart lighting is also important for the illumination of plants. Urban agriculture is the process by which fruits and vegetables for human consumption are grown in indoor greenhouse environments within densely populated cities [20]. The so-called vertical farms increase the crop yield per unit land area, relative to open, "horizontal farming," substituting energy from the sun with artificial lighting [17]. Additionally, production of food in population centers reduces transportation costs associated with importing food from rural areas. Using LEDs as the illumination source, the wavelength of emission can be tailored for specific plants with minimal waste heat generation and minimal attraction of pests [20]. Currently, Infineon is manufacturing and marketing smart LED systems specifically for urban agriculture [21].

3.2. Smart sensing

The rapid growth of urban areas has a direct impact on the environment, which in turn affects the health and well-being of urban inhabitants. Two important elements that are essentials to humans, and life in general, are air and water. Photonic-based sensors play an essential role in monitoring and controlling air and water pollution [22]. A number of sensors based on optical effects have been demonstrated for monitoring air and water quality. Herein, we provide a brief review. We note that smart water management generally was recently highlighted in [23]. Additionally, in a comprehensive review of advanced sensing networks for Smart Cities, Hancke et al. [24] identified use of optical fibers and lasers in structural health monitoring and electrical transmission line monitoring, respectively.

In the urban environment, determining the quality of air and water is a first step towards improving quality of life for inhabitants. Key analytes of the atmosphere include particulate matter, ground-level ozone, CO, NO_x, SO_2, and lead [25]. Key analytes of water include salinity, pH, chlorine, heavy metals, and bacteria. While necessary, simply determining concentrations of these analytes is not sufficient for reducing potential hazards for urban citizens. Smart Cities ought to strive to connect sensing modalities to integrated collection and decision-making operations to mitigate the source of contaminants in the urban water or oxygen supply [26]. Before discussing smart optical sensors networks, we provide a brief review of photonic technologies for sensing contaminants in air and water. Important parameters for evaluating the performance of a given sensor include sensitivity, selectivity, response time, reversibility, amount of collected information, power consumption, and cost. Sensitivity refers to the minimum quantity of detectable matter in a given volume of gas or water. Selectivity is the ability of the sensor to identify a specific element, molecule, or compound among a gaseous or liquid mixture. Response time can be very important in the cases where real-time detection and monitoring are mandatory, for instance, in areas that are needed to be kept safe from security threats. Also, the response time and amount of collected information have direct implication on the communication bandwidth, important for a network of sensors and actuators. Reversibility refers to whether the matter, upon being sensed, can return to its pre-sensed form. Power consumption and cost becomes especially important if a large number of sensors are implemented throughout a city.

Aside from sensors based on optical excitation, many sensors utilize the electrical response of materials, including metal oxides, semiconductors, and polymers. For instance, the detection of a specific gas can be sensed by measuring the variation in the conductance of a metal oxide [27]. Advantages of sensors based on electronic material properties include low cost and short response time. Disadvantages include low sensitivity, poor selectivity, irreversibility of some materials, and sensitivity to environmental factors, for example, temperature [28].

Most optical sensors are based on spectroscopy. Advantages of sensors based on optical response of materials include high sensitivity, high selectivity, high stability, long life time, short response time for real-time detection, and robustness to environmental factors. Disadvantages tend to include a large footprint and high cost [28].

Spectroscopic analysis mainly involves techniques based on absorption and emission spectrometry. Absorption spectroscopy, based on the Beer–Lambert law, is the concentration-dependent absorption of photons at specific wavelengths. The absorption spectra of specific gases can be found in the HITRAN database [29]. Some techniques based on absorption spectroscopy include differential optical absorption spectroscopy (DOAS) [30], tunable diode laser absorption spectroscopy (TDLAS) [31], light detection and ranging (LIDAR), Raman LIDAR [32], differential LIDAR (DIAL) [33], and intra-cavity absorption spectrometry (ICAS) [34]. These methods tend to be bulky and expensive. Furthermore, most of these techniques employ long wavelengths, requiring tens of meters of open space for operation. For applications in Smart City sensing, optical components must be miniaturized, to approach their practical microscale to nanoscale limit and thereby lead to compact absorption spectroscopy systems.

Fourier transform infrared spectrometry (FTIR) is a powerful technique with applications in environmental monitoring, including pollution at power plants, petrochemical and natural gas plants, waste disposals, agricultural, and industrial sites, and the detection of gases produced in flames, in biomass burning, and in flares [35]. The National Science Foundation's Center for Mid-Infrared Technologies for Health and the Environment (MIRTHE) has been using quantum cascade lasers (QCLs) to monitor air quality, including methane, ammonia, and other "molecular footprints" [36]. For example, the air quality of Beijing before, during, and after the 2008 Olympic Games was measured. QCLs also enable detection of natural-gas leaks and ground-based verification of remote sensors on aircraft and spacecraft. Thus far, measurements require large optical path lengths. Therefore, the technology is currently useful for inter-building distances in a city, but not at the intra-building scale [36].

In the intra-building scale, sensing of gases by photonic technologies was reviewed generally in [37]. Commercially available technologies include palm-size optical dust sensors [38] and dangerous-particle detectors [39]. Particulate matter less than 2.5 μm in diameter is especially prone to cause problems in urban environments. Such small particles come from industrial and automotive exhaust and often lead to cardiac and lung diseases [39]. In [38], an infrared emitting diode and a phototransistor are combined to enable detection of light scattered from airborne dust. The output of the sensor is an analog voltage proportional to the measured dust density, with a sensitivity of 0.5 V/0.1 mg/m^3. Shorter wavelengths could increase sensitivity [38]. Mitsubishi Electric recently claimed that a unique, double-sided mirror design is able to collect nearly twice as much scattered light as conventional single-sided designs for small-particle detection [39]. A proprietary algorithm is then able to distinguish between pollen and dust, based on the respective differences in the optical characteristics of their scattered light. The air-quality sensor prototype is based on a laser diode, aspheric lens, light-collecting mirror, photodetector, and airflow controller. The prototype measures just 67 x 49 x 35 mm, and the company says that it can detect particle sizes down to 0.3 μm in diameter ($PM_{2.5}$). The smaller $PM_{2.5}$ is produced by various combustion processes, including those in motor vehicles, power plants, and residential wood burners [39]. In [40], a sensor for sulfur dioxide was presented based on differential absorption spectroscopy of UV light. The device had a reported temporal resolution of 3 s, optical path length of 19.6 m, and minimum detection limit of 75 ppb SO_2. Applications for detecting levels of vehicle exhaust were discussed. In [41], sensing agents coated on the surface of bent optical fiber probes were analyzed for the detection of ammonia and carbon dioxide, and sensing humidity.

Concerning water sensing, recently, in [42] a real-time in-line bacteria sensor was demonstrated based on 3D image recognition. The device showed a 10-min temporal resolution and classified particulates as bacterial or abiotic based on over 50 image parameters. Field trials demonstrated that rapid changes in bacteria composition could be detected, in both pure and mixed solutions. In [43], the absorption of light in water was measured for wavelengths in the 300–800 nm range, as a function of both salinity and temperature. The resulting data provide a baseline for sensing local changes to water conditions. In [44], a water pollution monitoring system was demonstrated utilizing a water-core waveguide and UV illumination. Traces of nitrate and chlorine as low as 22 μg/L and 26 μg/L were detected, respectively. In [45], it was demonstrated that a cavity could significantly improve the absorption sensitivity of FTIR spectrometry by increasing the effective path length. This is important for achieving high sensitivity, while maintaining a small footprint for the detection of pollution in low-volume samples.

Aside from discrete devices, photonic sensors have benefited significantly from the maturity of CMOS technology, pushing on-chip integration toward microscale and nanoscale footprints. The emergence of CMOS sensors is attractive not only for the integration purposes but also due to incomparable advantages related to high-sensitivity, high-speed response, electromagnetic immunity, and low cost. Numerous high-performance integrated devices have been developed for chemical, biochemical, and gas detection. These sensors are based on various topologies that utilize changes to the local refractive index, including high-quality factor resonators [46, 47], Mach-Zehnder interferometers [48], 2D photonic crystal microcavities [49], and surface plasmon resonances [50, 51].

3.3. Smart communications and signal processing

Because of its unmatchable propagation velocity and carrier frequencies, light is the physical medium of choice through which information is carried over long distances [52]. Gradually, the meaning of "long distances" has evolved, as the miniaturization of optical components has enabled cost-effective optical links over smaller and smaller distances, while the demand for bandwidth has increased. As internet traffic becomes increasingly dominated by activity within data centers, the need for inexpensive, compact, fast, and efficient integrated optical components will increase. Fiber optic networks (FON) are already essential to the operation of cities, providing two-way communication between residents, businesses, and the rest of the world.

FON are smart systems that correct themselves when error or failures on the network arise. The transceivers used on FON contain detectors that monitor back-reflections from the fibers through optical-time domain reflectometry (OTDR) [53]. If there is a failure, back-reflections increase and an upper electronic layer is programmed to re-route the network, reducing errors and delays. We believe OTDR can inspire a new generation of self-controlled networks of sensors for efficient real-time decision making.

Fiber-to-the-home (FTTH) is the final leg of a FON and has become a critical component of Smart Cities [54]. Because a Smart City contains an extensive network of sensing and actuating nodes, fast and efficient communications between nodes are essential for effective operation. FTTH provides the fastest communication links currently conceivable and is only expected to become more widespread, interfacing with wireless communication and mobile platforms [54].

LEDs were mentioned previously as main components in smart lighting. Additionally, LEDs can simultaneously function as a communication channel, making a host of systems in the city smarter [55, 56]. Visible light communications (VLC) were first proposed decades ago for indoor communications [57], and experienced a resurgence of attention in 2004 [58]. Soon thereafter, the networking benefits of light-fidelity (Li-Fi) networks relative to RF-based Wi-Fi networks were demonstrated [59], followed by data rates exceeding 500 Mbps [60], and by a comprehensive analysis of the potential of VLC [56]. A multiplexed VLC system based on individual red, green, and blue LEDs exceeding a data rate of 3 Gbps was demonstrated in [61].

In addition to high-speed and high-capacity data transmission, data processing increasingly occurs in the optical domain [62, 63]. Photonic signal processing (PSP) enables multi-GHz sampling of RF or microwave signals, bypassing the inherent time-bandwidth limitations of electrical systems, with immunity to electromagnetic interference [64]. Widely tunable filters, waveform generators, Hilbert transformers, wave mixers, and signal correlators built from photonic devices have all been demonstrated, with inherent compatibility with fiber optics communications [62]. Furthermore, use of plasmonic materials suggests avenues for PSP elements in nanoscale footprints [65]. In the context of Smart Cities, compact PSP systems should ensure the processing of large amounts of information from arrays of sensors monitoring the urban environment.

4. Potential applications of photonics to Smart Cities

Having reviewed existing photonic technologies in the application areas of lighting, sensing, and communications and signal processing for Smart Cities, we now explore advanced applications. For concreteness, we use the recent water crisis in Flint, MI, USA, as well as the ongoing drought in San Diego, CA, USA, as real-world examples for proposing a smart network of integrated optical water resource sensors. We then discuss how Smart City concepts will drive long-term research goals in the photonics community.

Figure 3 summarizes a roadmap for photonic technologies for addressing urban water problems. Existing optics-based sensors have the ability to independently measure concentrations of pH, salinity, and bacteria, as well as the amount of water consumed. We propose to integrate these discrete sensors into a compact, cost-effective, energy-efficient system with graphical user interface (GUI) and connectivity to the outside world enabled by RF or optical transmitters (Tx) for wireless or plastic optical fiber (POF) communications. Finally, we anticipate that Smart City initiatives may drive research goals in photonics, including towards chip-scale spectrometers utilizing nanoscale light sources and metamaterials.

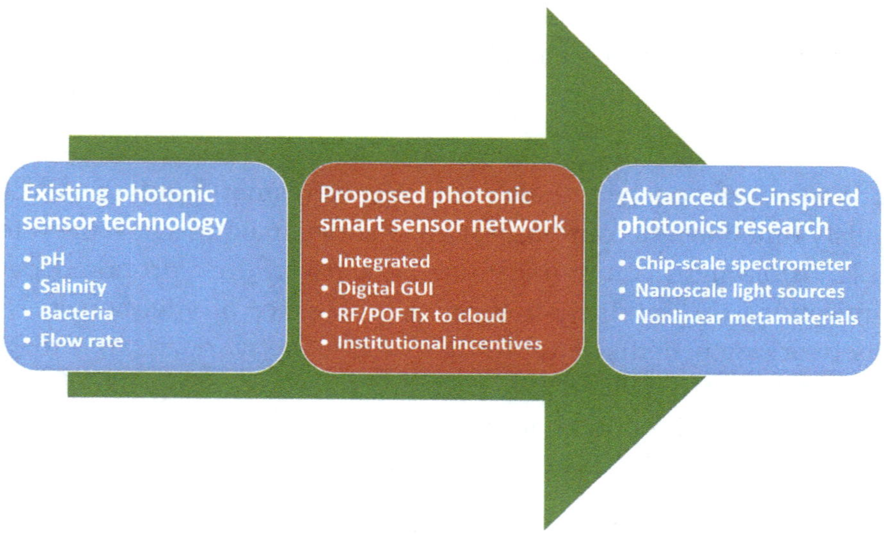

Figure 3. Technology roadmap of smart optical network for water-supply monitoring, highlighting existing sensing modalities, and potential intermediate, and long-term research topics.

4.1. Flint water pollution and the southern California drought

The recent drinking water crisis in Flint, Michigan, stands as a reminder of the fragility of civil infrastructure and poor decision making [66]. Century-old lead pipes leached lead into the drinking water supply after the source of the water was switched from Detroit to the Flint River. While elevated lead levels were noticed and reported by residents, local and state officials ignored changes to the water chemistry that caused leaching and eventual lead pollution [66]. Could this crisis, and situations like it, be avoided in the future with smart sensor networks? How can this crisis inspire future photonic technologies, in both the near and long terms?

Southern California has experienced an ongoing drought that affects over 34 million people (see **Figure 1a**) [67]. Because much of the population lives in arid regions, coastal cities in the region rely on imported water. This is exemplified by the water supply of San Diego, about

85% of which is imported from outside sources, such as the Colorado River and Sacramento Bay Delta (**Figure 4b**) [68]. Reduced water consumption has, therefore, become a major strategy for the region to retain a sufficient long-term supply [69]. Additionally, water from the Colorado River contains elements from old mining and industrial sites, while the water from the State Water Project contains traces of pesticides, herbicides, and high bromide levels. Efficient water-quality monitoring again becomes fundamental for the safety of urban dwellers [68]. How can smart sensor networks enable more efficient use of water? How can the longstanding drought conditions inspire future photonic technologies? How can smart sensors assist on water quality real-time monitoring?

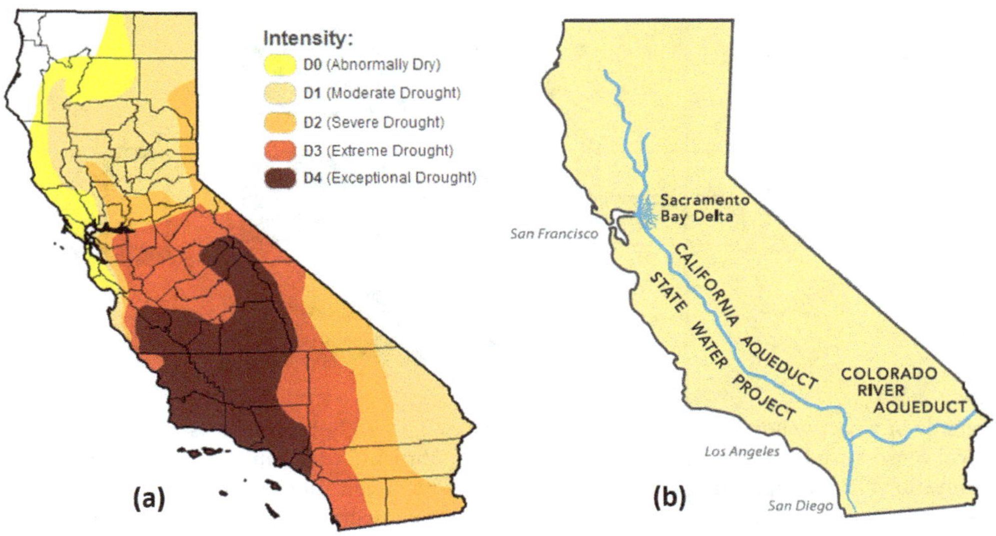

Figure 4. (a) Map of drought conditions in state of California as of May 24, 2016 [67]. (b) Map of California highlighting the main sources of water supplying the San Diego region [68].

4.2. Preventing future crises

To prevent the occurrence of pollution or resource-driven water crises, we propose an example of a smart water-supply system enabled by integrated optical sensors and communications. Recently, the city of Na Ding, Vietnam, modernized their water-supply system, installing sensors for salinity, pH, turbidity, and chlorine, to improve water treatment options [70]. Our proposed system goes further, detecting bacteria and consumption rates and relaying information directly to residents.

The proposed smart water system is shown in **Figure 5**. Integrated photonic sensors (red circles) simultaneously monitor pH, salinity, bacteria, and flow levels, and transmit this information to data aggregators and individual users. The transceiver architecture depends on data requirements, as outlined in [24]; we illustrate the possibility of wireless data transmission to residents and fiber links between regional level sensors and the aggregator. Note that wireless transmission may be either RF or VLC. Aggregated data are analyzed, creating information upon which decisions may be made, affecting the water supply and/or water treatment. All information is stored on the cloud for meta-analysis and eventual formation of knowledge, as advocated in [2].

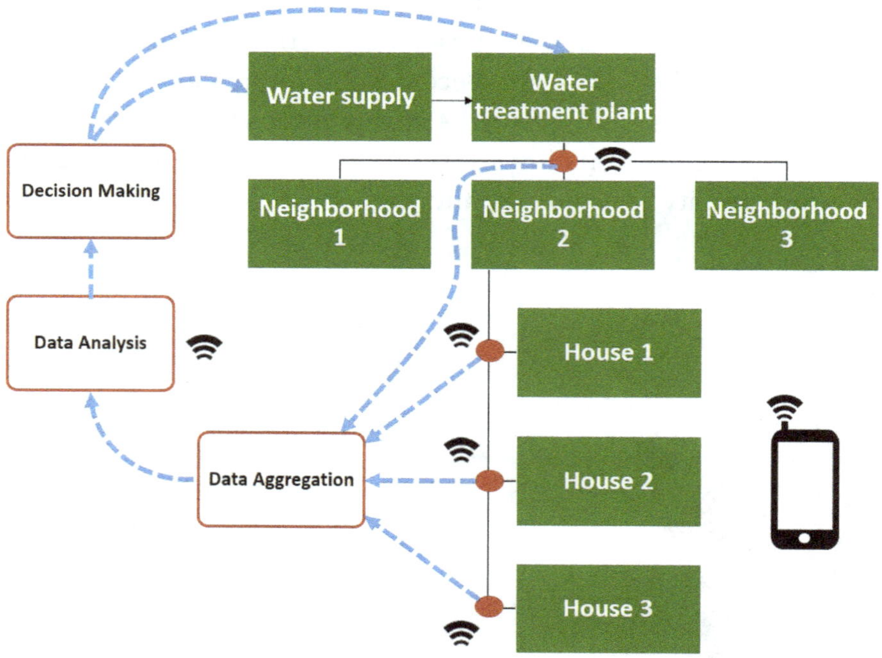

Figure 5. Schematic of smart water-supply system enabled by integrated optical sensors and optical communications. Integrated optical sensors (red circles) generate data on water quality and consumption. Data are transmitted by RF or optical communications links, depending on data rate requirements. Optical links transmit aggregated data for analysis. Decision is made based on analysis to actuate water supply or treatment plant appropriately, for example, alter salinity levels. Additionally, generated local data are transmitted to citizens via smart phone app and analyzed aggregate data are made accessible via public database.

Reflecting on the case of Flint, Michigan, the proposed smart water system could in principle prevent widespread pollution created by human error [66]. Firstly, the system could be implemented with autonomous decision making and actuation such that the water chemistry would be altered in response to measurements throughout the network. Through iterative sensing and actuation, lead levels would decrease in response to reaching a water chemistry wherein leaching would not occur. Secondly, real-time transmission of data to individual users and online databases would enable greater citizen participation in resource management. Consequently, local and state officials could be more easily held accountable for their actions or inactions in response to system problems.

For San Diego, California, the proposed smart water system could significantly reduce water consumption, a primary goal of the city's Climate Action Plan [69]. Transmission of real-time water use to customers would enable them to quantify wasteful practices. Employers could incentivize environmentally conscious behavior by awarding employees who make their water use data available and meet resource consumption goals [71]. A smart water system thereby enables the gamification of resource management, much like step-counters incentivize employees to practice preventive healthcare.

The smart water-supply system proposed here shares features with smart air-quality monitoring systems. For example, in Amsterdam, Netherlands, a start-up named TreeWiFi installed smart birdhouses to monitor the amount of combustion particles (NO_2) in the air [72]. LED lights placed on the roof of the birdhouse show real-time levels of pollution, and if the lights go green, which means improved quality of air, the network makes free Wi-Fi spots available. The next step is to make the collected data available to researchers, governmental departments, and the public, as we have also proposed in our smart water-supply system.

4.3. Advanced photonic technologies for Smart Cities

Sensing and monitoring systems for Smart Cities present a continuous cycle of operation: sensing—communication—decision making—sensing. Returning to **Figure 1**, Smart City requirements will inspire the development of advanced photonic technologies such as detectors and sensors, light sources, modulators, and optical hardware accelerators offering unprecedented speeds for communication and decision making, while consuming low power in a small footprint. Based on these requirements, we briefly describe our vision on how Smart Cities can drive research in advanced nanophotonic technologies.

In the example of the birdhouses, the sensors are purely electronic, presenting a large footprint for detecting just one type of molecule. However, in Section 3(b), we have shown examples of water-quality monitoring through absorption of light in water, where different choice of wavelengths allows probing different properties of water, such as temperature, salinity, pH, and traces of nitrates and chlorine. In the same way, different wavelengths can monitor different molecular constituents of air. Photonics then brings a new concept to sensing, sensor fusion. In data processing, this concept is related to combining sensory data from different sources to reduce the uncertainty in the resulting data. Here, we extend the concept to a sensor that simultaneously probes different properties of the desired environment. An array of five semiconductor nanolasers [73, 74], each one with all spatial dimensions less than 1 μm, placed 1 μm from each other, could provide light emission in five different wavelengths on an array pitch of 10 μm. A semiconductor laser, if reversed biased, can act as photodetector, which means a similar compact array could be used for power detection. Inserting now a medium to be monitored between these two arrays, one can sense five different properties of the medium, where each property is addressed by one wavelength. Recent advances to increase the efficiency of these semiconductor nanolasers operating at room temperatures will make this technology available in the near future [73].

After detection, a communication channel is necessary. Here, current optical communications technologies can play a major role from which we can learn. Previously, we explained the concept of FTTH, where the objective is replacing existing copper infrastructure for telecommunications by optical fibers, providing vastly higher bandwidths and enabling more robust internet services to the end consumer. We envision an active optical network that we name fiber-to-the-sensors (FTTS). All fused sensors, in the near future, are connected by optical fibers using the same protocols and fiber optic cable infrastructure already used on communication systems. In FTTH, information from different users can use different channels (frequencies), which are multiplexed, transmitted across long distances, and then demultiplexed to reach the final users. On FTTS, information of different frequencies (measured water or air properties) are multiplexed and transmitted with fiber optics to a central node, where information is demultiplexed to be classified and used on a decision-making process. The information on quality and quantity consumed is also returned to the user. Multiplexers [75], demultiplexers, efficient laser sources for transmitters [76], efficient and sensitive detectors for receivers, fast switches and routers based on nonlinear optical process [77], and other photonics technologies are needed to increase the bandwidth and reduce power consumption in optical interconnects that allows FTTH, and possibly FTTS [78]. If the number of sensors in a Smart City starts to increase, fast communication, and data processing is necessary for fast decision making. In this case, all technology that has been developed for fast, robust, and low power consumption optical interconnects in data centers can be applied on a FTTS network [73, 79, 80]. Here, central nodes that collect information from different sensors are the data centers. Furthermore, OTDR systems used for fiber fault detection can be applied to detect sensor failures and reroute the network.

Besides smart sensors networks, there are several other Smart City needs than can benefit from photonics. Continued progress is needed in the area of mid- and far-infrared photonics for developing optical sources emitting in atypical frequencies, such as within the Terahertz window. Working in these frequencies allows monitoring optical absorption of elements that cannot presently be monitored, resulting in ubiquitous sensing capabilities for monitoring air and water pollution. Consequently, detection in these frequencies regimes will also be necessary. One of the candidates for enhanced emission and detection in these regimes is III–V semiconductors coupled to plasmonic inclusions [81–83]. The first luminescent hyperbolic metamaterial was developed recently by our group, operating at the C telecommunication band [84–86]. However, the capability of tuning the constituent materials allows, in principle, absorption and emission in the required atypical wavelengths. Other application of metamaterials includes perfect absorbers for solar cells that can enhance the energy harvesting and super-lenses for imaging with higher resolutions [87].

More progress is also needed in the engineering of light-matter interactions, for increasing the fundamental speed limit and efficiencies with which optoelectronic devices may be modulated [88]. While devices based on stimulated emission, for example, semiconductor lasers, are limited to less than 100 GHz modulation bandwidth due to relaxation oscillations, it is conceivable to design faster devices based on spontaneous emission [89]. No devices have yet approached the fundamental limits to the enhancement of linear radiative processes [90] or nonlinear processes [91], necessitating that engineered nanostructures and metamaterials remain an active focus of research for the benefit of all Smart City applications.

4.4. Mitigating deleterious effects of future crises

Finally, efficient decision making is extremely necessary, as we can remember from the Flint case; a human decision making, or rather negligence, led to a catastrophic scenario. In the last few years, there is a trend on researching optical hardware accelerators to solve large-scale, high-complexity problems using brain inspired architectures for efficient pattern recognition with low power consumption. Examples of proposed and demonstrated devices include nonlinear coupled semiconductor lasers for pattern recognition and decision making [5] and photonic time stretching for processing images and information [92]. It is expected that analog computers and hardware accelerators will advance significantly in the next few years. While the current focus is to assist electronic-based computation, there is plenty of space for developing new architectures designed for Smart Cities.

As an example, we consider the application of optical computing to problems in infrastructure resiliency. The resiliency of a system is a measure of its ability to withstand external forces, respond quickly to damages, and return to a normal state of operation [93]. Negative external forces include natural disasters, industrial accidents, and terrorist attacks. Quantitative descriptions of resiliency provide a means for optimizing the system by minimizing the costs associated with disruptions and system downtime [94]. These optimization problems are often formalized in terms of a multimode resource-constrained project scheduling problem [95], which may be solved using the simulated annealing algorithm [96, 97].

Physical implementation of simulated annealing with optical components was first investigated decades ago [98, 99]. Recently, optical hardware accelerators have gained more traction for assisting electronic digital information processing, for particular classes of problems wherein the difficulty scales nonlinearly with problem size, such as modeling metastable heteroclinic channels [100]. An initial concept with this focus in mind was recently proposed in [5]. We believe that continued co-optimization of nanophotonic materials, devices, and system architectures could make photonic hardware accelerators competitive for solving urban management problems of the future.

5. Conclusion

We have identified the major applications of photonics to Smart Cities and outlined topical areas for future research. It is hoped that this chapter serves simultaneously as a review of the impact of photonics on Smart Cities and as a roadmap for photonics research inspired by the demands of Smart Cities. As the global population becomes increasingly urban, photonics-based solutions are increasingly needed for improving the lives of all urban-dwellers.

Acknowledgements

This work was supported by the Office of Naval Research Multidisciplinary Research Initiative (N00014-13-1-0678), the National Science Foundation (NSF) (ECE3972 and ECCS-1229677), the NSF Center for Integrated Access Networks (EEC-0812072, Sub 502629), and the Cymer Corporation.

Author details

Joseph S.T. Smalley*, Felipe Vallini, Abdelkrim El Amili and Yeshaiahu Fainman

*Address all correspondence to: jsmalley@ucsd.edu

Department of Electrical & Computer Engineering, University of California, San Diego, La Jolla, CA, USA

References

[1] R. Glass, A. L. Ames, W. A. Stubblefield, S. H. Conrad, S. L. Maffitt, L. A. Malczynski, D. G. Wilson, J. J. Carlson, G. A. Backus, M. A. Ehlen, K. B. Vanderveen, and D. Engi, A Roadmap for the Complex Adaptive Systems of Systems (CASoS) Engineering Initiative, Sandia National Laboratory, Albuquerque, NM (2008).

[2] A. Ramaswami, A. G. Russell, P. J. Culligan, K. R. Sharma, and E. Kumar, "Meta-principles for developing smart, sustainable, and healthy cities," Science 352, 940–943 (2016).

[3] M. Mitchell, Complexity: A Guided Tour, Oxford University, New York (2009).

[4] A. Zanella, N. Bui, A. Castellani, L. Vangelista, and M. Zorzi, "Internet of things for smart cities," IEEE Internet Things J. 1, 22–32 (2014).

[5] S. Shahin, F. Vallini, F. Monifi, M. Rabinovich, and Y. Fainman, "Heteroclinic dynamics in photonic cognitive motif network," in Conference on Lasers and Electro-Optics, OSA (2016), p. ATu3O.3.

[6] H. Chourabi, J. R. Gil-Garcia, T. A. Pardo, T. Nam, S. Mellouli, H. J. Scholl, S. Walker, and K. Nahon, "Understanding smart cities: An integrative framework," in 45th Hawaii International Conference on System Sciences, Honolulu, HI (2012), pp. 2289–2297.

[7] C. Harrison, B. Eckman, R. Hamilton, P. Hartswick, J. Kalagnanam, J. Paraszczak, and P. Williams, "Foundations for smarter cities," IBM J. Res. Dev. 54, (2010).

[8] S. Dirks and M. Keeling, A Vision of Smarter Cities: How Cities Can Lead the Way into a Prosperous and Sustainable Future, IBM Institute for Business Value, North Castle, NY (2009).

[9] L. Bettencourt, J. Lobo, D. Helbing, C. Kuhnert, and G. West, "Growth, innovation, scaling, and the pace of life in cities," Proc. Natl. Acad. Sci. 104, 7301–7306 (2007).

[10] L. Bettencourt, "The origins of scaling in cities," Science 340, 1438–1441 (2013).

[11] How to Transform a City: Lessons from the IBM Smarter Cities Challenge, IBM Smarter Cities White Paper (2012).

[12] How to Reinvent a City: Mayors' Lessons from the Smarter Cities Challenge, IBM Smarter Cities White Paper (2013).

[13] "Key technology manufacturing areas," http://www.aimphotonics.com/key-technology-manufacturing-areas/, American Institute for Manufacturing, Albany, NY (2016).

[14] A. Sabapathy and J. Whittaker, "An Assessment of Energy Efficiency and Carbon Reductions from Replacing Halogen Lamps with Light Emitting Diodes in Traffic Lights – A Case Study of Gloucestershire County's Initiative," Energy Environ. 22, 361–374 (2011).

[15] J. Lai, J. Qiu, J. Chen, Y. Wang, and H. Fan, "Application of wireless intelligent control system for HPS lamps and LEDs combined illumination in road tunnel," Comput. Intell. Nueroscience 2014, 429657 (2014).

[16] S. Walerczyk, "Human-centric lighting," http://humancentriclighting.org/, Lighting Wizards (2012).

[17] D. Despommier, The Vertical Farm: Feeding the World in the 21st Century, Macmillan, New York (2011).

[18] M. S. Mott, D. H. Robinson, A. Walden, J. Burnette, and A. S. Rutherford, "Illuminating the effects of dynamic lighting on student learning," SAGE Open, http://dx.doi.org/10.1177/2158244012445585 (2012).

[19] T. Herbst, "Why lighting systems will become more connected," in 2015 Connected Lighting Systems Meeting, U.S. Department of Energy, Washington, D.C. (2015).

[20] M. Rozenfeld, "Indoor farms could revolutionize agriculture: No soil, sunlight, or pesticides required," IEEE Institute, http://theinstitute.ieee.org/technology-topics/life-sciences/indoor-farms-could-revolutionize-agriculture (Published online May 23, 2016).

[21] Infineon, "LEDs in urban farming," https://www.infineon.com/cms/en/about-infineon/company/our-contribution/urbanfarming/, Infineon (2016).

[22] T. Kjellstrom, M. Lodh, T. McMichael, G. Ranmuthugala, R. Shrestha, and S. Kingsland, Air and Water Pollution: Burden and Strategies for Control, The International Bank for Reconstruction and Development/The World Bank, Washington, D.C. (2006).

[23] T. A. Larsen, S. Hoffmann, C. Lüthi, B. Truffer, and M. Maurer, "Emerging solutions to the water challenges of an urbanizing world," Science 352, 928–933 (2016).

[24] G. P. Hancke and B. de C. E. Silva, "The role of advanced sensing in smart cities," Sensors (Basel). 13, 393–425 (2013).

[25] S. Devarakonda, P. Sevusu, H. Liu, R. Liu, L. Iftode, and B. Nath, "Real-time air quality monitoring through mobile sensing in metropolitan areas," in Proceedings of the 2nd

ACM SIGKDD International Workshop on Urban Computing – UrbComp'13, ACM Press, Chicago, IL (2013), p. 1.

[26] K. Pretz, "An urban reality: Smart cities," IEEE Institute, http://theinstitute.ieee.org/technology-focus/technology-topic/an-urban-reality-smart-cities (Published online 2014).

[27] A. Fioravanti, A. Bonanno, S. Gherardi, M. C. Carotta, and A. N. Skouloudis, "A portable air-quality station based on thick film gas sensors for real time detection of traces of atmospheric pollutants," IOP Conf. Ser. Mater. Sci. Eng. 108, 12005 (2016).

[28] X. Liu, S. Cheng, H. Liu, S. Hu, D. Zhang, and H. Ning, "A survey on gas sensing technology," Sensors 12, 9635–9665 (2012).

[29] L. S. Rothman, I. E. Gordon, Y. Babikov, A. Barbe, D. Chris Benner, P. F. Bernath, M. Birk, L. Bizzocchi, V. Boudon, L. R. Brown, A. Campargue, K. Chance, E. A. Cohen, L. H. Coudert, V. M. Devi, B. J. Drouin, A. Fayt, J.-M. Flaud, R. R. Gamache, J. J. Harrison, J.-M. Hartmann, C. Hill, J. T. Hodges, D. Jacquemart, A. Jolly, J. Lamouroux, R. J. Le Roy, G. Li, D. A. Long, O. M. Lyulin, C. J. Mackie, S. T. Massie, S. Mikhailenko, H. S. P. Müller, O. V. Naumenko, A. V. Nikitin, J. Orphal, V. Perevalov, A. Perrin, E. R. Polovtseva, C. Richard, M. A. H. Smith, E. Starikova, K. Sung, S. Tashkun, J. Tennyson, G. C. Toon, V. G. Tyuterev, and G. Wagner, "The HITRAN2012 molecular spectroscopic database," J. Quant. Spectrosc. Radiat. Transf. 130, 4–50 (2013).

[30] Z. Haiming, "Experiment study of continuous emission monitoring system based on differential optical absorption spectroscopy," in 2008 International Workshop on Education Technology and Training & 2008 International Workshop on Geoscience and Remote Sensing, IEEE, Beiging, China (2008), Vol. 1, pp. 175–177.

[31] Y. Miao, Q. Yao, N. Qiu, and J. Zhang, "Application research of laser gas detection technology in the analysis of sulphur hexafluoride," CICED 2010 Proc. 1–3, Nanjing, China (2010).

[32] H. Miya, T. Shiina, T. Kato, K. Noguchi, T. Fukuchi, I. Asahi, S. Sugimoto, H. Ninomiya, and Y. Shimamoto, "Compact Raman lidar for hydrogen gas leak detection," in 2009 Conference on Lasers & Electro Optics & The Pacific Rim Conference on Lasers and Electro-Optics, IEEE, Shanghai (2009), pp. 1–2.

[33] S. Kuang, J. F. Burris, M. J. Newchurch, S. Johnson, and S. Long, "Differential absorption lidar to measure subhourly variation of tropospheric ozone profiles," IEEE Trans. Geosci. Remote Sens. 49, 557–571 (2011).

[34] K. Liu, T. Liu, J. Jiang, G. D. Peng, H. Zhang, D. Jia, Y. Wang, W. Jing, and Y. Zhang, "Investigation of wavelength modulation and wavelength sweep techniques in intracavity fiber laser for gas detection," J. Light. Technol. 29, 15–21 (2011).

[35] Z. Bacsik, J. Mink, and G. Keresztury, "FTIR spectroscopy of the atmosphere part 2. Applications," Appl. Spectrosc. Rev. 40, 327–390 (2005).

[36] P. Daukantas, "Air-quality monitoring in the mid-infrared," Opt. Photonics News 26(11), 26–33 (2015).

[37] J. Hodgkinson and R. P. Tatam, "Optical gas sensing: a review," Meas. Sci. Technol. 24, 12004 (2013).

[38] "Optical dust sensor – GP2Y1010AU0F," Sparkfun product sheet (2016), https://www.sparkfun.com/products/9689.

[39] "Mitsubishi air quality sensor detects dangerous particulates," Optics.org (Published online February 9, 2016), http://optics.org/news/7/2/13.

[40] J. T. Pisano, C. G. Sauer, J. Robbins, J. W. Miller, H. Gamble, and T. D. Durbin, "A UV differential optical absorption spectrometer for the measurement of sulfur dioxide emissions from vehicles," Meas. Sci. Technol. 14, 2089–2095 (2003).

[41] L. Xu, J. C. Fanguy, K. Soni, and S. Tao, "Optical fiber humidity sensor based on evanescent-wave scattering," Opt. Lett. 29, 1191 (2004).

[42] B. Højris, S. C. B. Christensen, H.-J. Albrechtsen, C. Smith, and M. Dahlqvist, "A novel, optical, on-line bacteria sensor for monitoring drinking water quality," Sci. Rep. 6, 23935 (2016).

[43] W. S. Pegau, D. Gray, and J. R. V. Zaneveld, "Absorption and attenuation of visible and near-infrared light in water: dependence on temperature and salinity," Appl. Opt. 36, 6035 (1997).

[44] P. Dress, M. Belz, K. F. Klein, K. T. V. Grattan, and H. Franke, "Water-core waveguide for pollution measurements in the deep ultraviolet," Appl. Opt. 37, 4991 (1998).

[45] A. A. Ruth, J. Orphal, and S. E. Fiedler, "Fourier-transform cavity-enhanced absorption spectroscopy using an incoherent broadband light source," Appl. Opt. 46, 3611 (2007).

[46] X. Fan, I. M. White, H. Zhu, J. D. Suter, and H. Oveys, "Overview of novel integrated optical ring resonator bio/chemical sensors," in A. V. Kudryashov, A. H. Paxton, and V. S. Ilchenko, eds., International Society for Optics and Photonics, San Jose, CA (2007), p. 64520M.

[47] H. K. Hunt and A. M. Armani, "Label-free biological and chemical sensors," Nanoscale 2, 1544 (2010).

[48] F. Prieto, B. Sepulveda, A. Calle, A. Llobera, C. Dominguez, A. Abad, A. Montoya, and L. M. Lechuga, "An integrated optical interferometric nanodevice based on silicon technology for biosensor applications," Nanotechnology 14, 907–912 (2003).

[49] B. Troia, A. Paolicelli, F. De, and V. M. N. Passaro, "Photonic crystals for optical sensing: A review," in Advances in Photonic Crystals, InTech, Rijeka, Croatia (2013).

[50] J. Homola, "Surface plasmon resonance sensors for detection of chemical and biological species," Chem. Rev. 108, 462–493 (2008).

[51] B. Sepúlveda, P. C. Angelomé, L. M. Lechuga, and L. M. Liz-Marzán, "LSPR-based nanobiosensors," Nano Today 4, 244–251 (2009).

[52] G. Agrawal, Nonlinear Fiber Optics, Academic Press, San Diego, CA (2007).

[53] M. K. Barnoski, M. D. Rourke, S. M. Jensen, and R. T. Melville, "Optical time domain reflectometer," Appl. Opt. 16, 2375 (1977).

[54] R. Jones, "FTTH at the heart of a smart city," white paper, Fiber to the home council of Middle East and North Africa, Amman, Jordan (2015).

[55] H. Haas, "High-speed wireless networking using visible light," SPIE Newsroom (Published online April 19, 2013), DOI: 10.1117/2.1201304.004773.

[56] P. H. Pathak, X. Feng, P. Hu, and P. Mohapatra, "Visible light communication, networking, and sensing: A survey, potential and challenges," IEEE Commun. Surv. Tutorials 17, 2047–2077 (2015).

[57] F. Gfeller and U. Bapst, "Wireless in-house data communication via diffuse infrared radiation," Proc IEEE 67, 1474–1486 (1979).

[58] T. Komine and M. Nakagawa, "Fundamental analysis for visible-light communication system using LED lights," IEEE Trans. Consum. Electron. 50, 100–107 (2004).

[59] M. Z. Afgani, H. Haas, H. Elgala, and D. Knipp, "Visible light communication using OFDM," in 2nd International Conference on Testbeds and Research Infrastructures for the Development of Networks and Communities, 2006. TRIDENTCOM 2006, IEEE, Barcelona, Spain (2006), pp. 129–134.

[60] J. Vucic, C. Kottke, S. Nerreter, K.-D. Langer, and J. W. Walewski, "513 Mbit/s visible light communications link based on DMT-modulation of a white LED," J. Light. Technol. 28, 3512–3518 (2010).

[61] F. M. Wu, C. T. Lin, C. C. Wei, C. W. Chen, Z. Y. Chen, and H. T. Huang, "Performance comparison of OFDM signal and CAP signal over high capacity RGB-LED-based WDM visible light communication," IEEE Photonics J. 5, 7901507–7901507 (2013).

[62] R. A. Minasian, E. H. W. Chan, and X. Yi, "Microwave photonic signal processing," Opt. Express 21, 22918 (2013).

[63] J. Capmany, J. Mora, I. Gasulla, J. Sancho, J. Lloret, and S. Sales, "Microwave photonic signal processing," J. Light. Technol. 31, 571–586 (2013).

[64] R. A. Minasian, "Photonic signal processing of microwave signals," IEEE Trans. Microw. Theory Tech. 54, 832–846 (2006).

[65] A. V. Krasavin and A. V. Zayats, "Photonic signal processing on electronic scales: Electro-optical field-effect nanoplasmonic modulator," Phys. Rev. Lett. 109, 53901 (2012).

[66] M. Torrice, "How lead ended up in Flint's tap water," Chem. Eng. News 94, 26–29 (2016).

[67] D. Simeral, "United States drought monitor: California," http://droughtmonitor.unl.edu/Home/StateDroughtMonitor.aspx?CA, The National Drought Mitigation Center, Lincoln, NE (Published online May 24, 2016).

[68] E. Symons, Municipal report, City of San Diego 2014 Annual Drinking Water Quality Report, San Diego, CA (2015).

[69] F. Faulconer, Municipal report, City of San Diego Climate Action Plan, San Diego, CA (2015).

[70] J. Andrews, "Da Nang opens smart control centre," Cities Today, http://cities-today.com/da-nang-opens-smart-control-centre/ (Published online August 16, 2013).

[71] L. Galli, P. Fraternali, C. Pasini, G. Baroffio, A. Dos Santos, R. Acerbis, and V. Riva, "A gamification framework for customer engagement and sustainable water usage promotion," in 36th IAHR World Congress, Delft, Holland (2015).

[72] "In Amsterdam, clean air gets free wifi from trees," Springwise Intelligence Ltd., http://www.springwise.com/amsterdam-clean-air-free-wifi-trees/, (Published online June 3, 2016).

[73] Q. Gu, J. S. T. Smalley, M. P. Nezhad, A. Simic, J. H. Lee, M. Katz, O. Bondarenko, B. Slutsky, A. Mizrahi, V. Lomakin, and Y. Fainman, "Subwavelength semiconductor lasers for dense chip-scale integration," Adv. Opt. Photonics 6, 1 (2014).

[74] Q. Gu, J. S. T. Smalley, J. Shane, O. Bondarenko, and Y. Fainman, "Temperature effects in metal-clad semiconductor nanolasers," Nanophotonics 4, 26–43 (2015).

[75] D. T. H. Tan, A. Grieco, and Y. Fainman, "Towards 100 channel dense wavelength division multiplexing with 100GHz spacing on silicon," Opt. Express 22, 10408 (2014).

[76] O. Bondarenko, C.-Y. Fang, F. Vallini, J. S. T. Smalley, and Y. Fainman, "Extremely compact hybrid III-V/SOI lasers: design and fabrication approaches," Opt. Express 23, 2696 (2015).

[77] M. Puckett, J. Smalley, M. Abashin, A. Grieco, and Y. Fainman, "Tensor of the second-order nonlinear susceptibility in asymmetrically strained silicon waveguides: analysis and experimental validation," Opt. Lett. 39, 1693–1696 (2014).

[78] Y. Fainman, M. P. Nezhad, D. T. H. Tan, K. Ikeda, O. Bondarenko, and A. Grieco, "Silicon nanophotonic devices for chip-scale optical communication applications [Invited]," Appl. Opt. 52, 613 (2013).

[79] Y. Fainman and G. Porter, "Directing data center traffic," Science 342, 202–203 (2013).

[80] A. Grieco, G. Porter, and Y. Fainman, "Integrated space-division multiplexer for application to data center networks," IEEE J. Sel. Top. Quantum Electron. 22, 1–6 (2016).

[81] J. Lee, M. Tymchenko, C. Argyropoulos, P. Chen, F. Lu, F. Demmerle, G. Boehm, M. Amann, A. Alu, and M. Belkin, "Giant nonlinear response from plasmonic metasurfaces coupled to intersubband transitions," Nature 511, 65–69 (2014).

[82] O. Wolf, S. Campione, A. Benz, A. Ravikumar, S. Liu, T. Luk, E. Kadlec, E. Shaner, J. Klem, M. Sinclair, and I. Brener, "Phased-array sources based on nonlinear metamaterial nanocavities," Nat. Comm. 6, 7667 (2015).

[83] L. M. Freeman, A. Smolyaninov, L. Pang, and Y. Fainman, "Simulated Raman correlation spectroscopy for quantifying nucleic acid-silver composites," Sci. Rep. 6, 23535 (2016).

[84] J. Smalley, F. Vallini, B. Kante, and Y. Fainman, "Modal amplification in active waveguides with hyperbolic dispersion at telecommunication frequencies," Opt. Exp. 22, 21088–21105 (2014).

[85] J. Smalley, F. Vallini, S. Shahin, B. Kante, and Y. Fainman, "Gain-enhanced high-k transmission through metal-semiconductor hyperbolic metamaterials," Opt. Mat. Express 5, 2300–2312 (2015).

[86] J. S. T. Smalley, F. Vallini, S. Montoya, E. E. Fullerton, and Y. Fainman, "Practical realization of deeply subwavelength metal-dielectric nanostructures based on InGaAsP," in Proc. SPIE 9544, Metamaterials, Metadevices, and Metasystems 2015, San Diego, CA (2015).

[87] L. Ferrari, C. Wu, D. Lepage, X. Zhang, and Z. Liu, "Hyperbolic metamaterials and their applications," Prog. Quantum Electron. 40, 1–40 (2015).

[88] J. B. Khurgin and G. Sun, "Comparative analysis of spasers, vertical-cavity surface-emitting lasers and surface-plasmon-emitting diodes," Nat. Photonics 8, 468–473 (2014).

[89] M. S. Eggleston, K. Messer, L. Zhang, E. Yablonovitch, and M. C. Wu, "Optical antenna enhanced spontaneous emission.," Proc. Natl. Acad. Sci. U. S. A. 112, 1704–1709 (2015).

[90] O. D. Miller, A. G. Polimeridis, M. T. Homer Reid, C. W. Hsu, B. G. DeLacy, J. D. Joannopoulos, M. Soljačić, and S. G. Johnson, "Fundamental limits to optical response in absorptive systems," Opt. Express 24, 3329 (2016).

[91] M. Kuzyk, "Fundamental limits of all nonlinear-optical phenomena that are representable by a second-order nonlinear susceptibility," J. Chem. Phys. 125, 154108 (2006).

[92] C. L. Chen, A. Mahjoubfar, and B. Jalali, "Optical data compression in time stretch imaging," PLoS One 10, e0125106 (2015).

[93] C. S. Holling, "Resilience and stability of ecological systems," Annu. Rev. Ecol. Syst. 4, 1–23 (1973).

[94] E. D. Vugrin, M. A. Turnquist, and N. J. K. Brown, "Optimal recovery sequencing for enhanced resilience and service restoration in transportation networks," Int. J. Crit. Infrastructures 10, 218–246 (2014).

[95] F. F. Boctor, "Resource-constrained project scheduling by simulated annealing," Int. J. Prod. Res. 34, 2335–2351 (1996).

[96] S. Kirkpatrick, C. D. Gelatt, and M. P. Vecchi, "Optimization by simulated annealing.," Science 220, 671–80 (1983).

[97] V. Černý, "Thermodynamical approach to the traveling salesman problem: An efficient simulation algorithm," J. Optim. Theory Appl. 45, 41–51 (1985).

[98] A. J. Ticknor, Optical Computing in Boltzmann Machines, University of Arizona (1987).

[99] N. Farhat, "Optical computing based on neuronal models," Report for Program Review, University of Pennsylvania (1988).

[100] M. Rabinovich and P. Varona, "Transient brain dynamics," in Principles of Brain Dynamics: Global State Interactions, M. Rabinovich, K. Friston, and P. Varona, eds., MIT, Cambridge, MA (2012), pp. 71–92.

8

Computational Tools for Data Processing in Smart Cities

Danilo Hernane Spatti and
Luisa Helena Bartocci Liboni

Abstract

Smart Grids provide many benefits for society. Reliability, observability across the energy distribution system and the exchange of information between devices are just some of the features that make Smart Grids so attractive. One of the main products of a Smart Grid is to data. The amount of data available nowadays increases fast and carries several kinds of information. Smart metres allow engineers to perform multiple measurements and analyse such data. For example, information about consumption, power quality and digital protection, among others, can be extracted. However, the main challenge in extracting information from data arises from the data quality. In fact, many sectors of the society can benefit from such data. Hence, this information needs to be properly stored and readily available. In this chapter, we will address the main concepts involving Technology Information, Data Mining, Big Data and clustering for deploying information on Smart Grids.

Keywords: Big Data, Data Mining, clustering

1. Introduction to Big Data concepts

Since the 1970s, companies search for efficient schemes to use their data. In this area of knowledge, concepts of database systems are very common, with the large variety storage suppliers, companies that need large storage can make use of tools that best meet their needs without the premise of getting attached to a particular technology.

Database management systems (DMSs) integrate information systems of different companies, always efficiently making information available to users. In a contemporary context, we have seen a substantial growth of storage needs, particularly in the electric sector.

Also, it is possible to highlight the cheapening of storage technologies as well as the large offer of high-quality, decentralized professional services, such as those supported by cloud computing.

Therefore, with higher storage capacity and with greater demands for storing and registering all possible information for further analysis, the concept of Big Data emerges.

With increased volume of information, it becomes impossible to manage the database through conventional methods. The term Big Data stands for data that require special tools for information extraction since information of interest is immersed in a completely unobservable environment. A very simple example is shown in **Figure 1**.

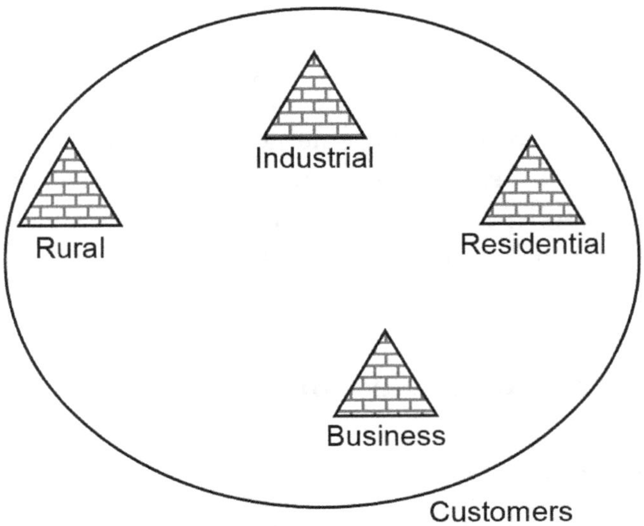

Figure 1. Example of databases.

In a set of customers stored in a naturalistic way, it is not possible to distinguish each of the internal components. In this case, the customer base has become a Big Data. Therefore, it is necessary to search for tools and construction rules that allow the correct distinction between the internal elements.

According to Manyika et al. [1], the implementation of information within the so-called Big Data environments account for a considerable share of the Information Technology market, with potential growth for years to come. To better understand the concept, four basic characteristics associated with Big Data systems can be defined:

1.1. Data volume

Consists in specifying the density of information that a database must support. Notably, we can highlight this as the most important characteristic in the DMS project. It depends on the nature of the information to be stored and also the type of communication channel to be used. The volume of Big Data systems is considered large and with exponential scalability.

1.2. Access speed

A robust DMS, able to store a multitude of data entities, must also possess a compatible access response. The evolution of database technology allowed the suppression of slow access, but this is still a crucial issue when specifying a DMS, which will be used in Big Data.

1.3. Data diversity

Big Data not only relates to pure information stored in tables. In some scenarios, DMS should be able to operate in environments containing several layers of information such as device status and keys, records, oscillography, spreadsheets, text documents, images, videos, etc. This is a prospect in the Smart Grid environment, considering that the integration of information is the basis for its operation.

1.4. Data consistency

A non-reliable database is a problem that emerges from data analysis. The out-of-control growth of the population of entities in a DMS can lead to situations where information quality is jeopardized. In this case, indispensable tools must ensure the integrity and reliability of Big Data databases, considering that operations are not possible with naturalistic observation.

2. Data Mining and attributes selection

2.1. Knowledge database discovery

Knowledge discovery in a database (KDD) is a field in Information Technology that integrates tools for automatically and intelligently analyse large repositories of information. It is an iterative process, whose main task is Data Mining [2]. **Figure 2** shows, in a simplified way, the process of KDD.

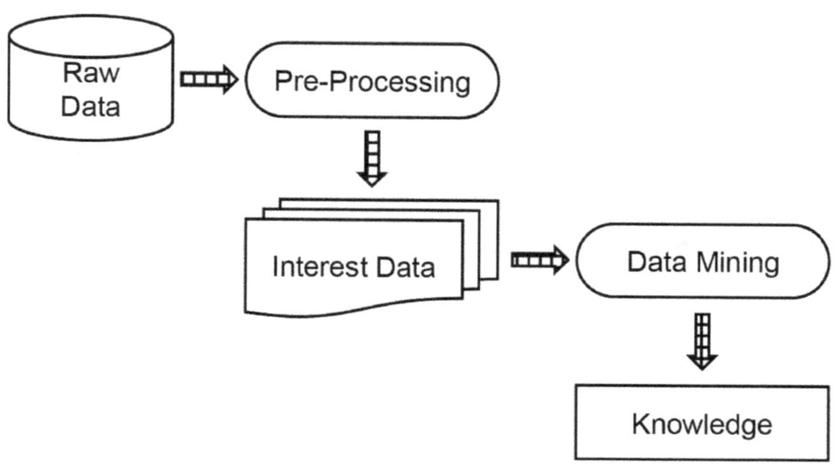

Figure 2. Simplified KDD process.

Data Mining is inserted in this context as one of the stages of KDD, responsible for metrics and methods that will extract the characteristics of interest. We can also highlight the pre-processing of raw data as another key step to reaching information knowledge.

It can be seen in **Figure 2** that the main concept involving Data Mining is to reduce the search space, resulting in a subset of data that is significantly smaller, but that substantially represents the main data set up [3].

KDD can play key roles in various industries, such as searches on financial information for loans, diagnosis for preventive maintenance systems, pattern detection in satellite images, among others [4]. One of the main units handled in KDD processes are the attributes, which are responsible for characterizing information.

2.2. Attribute selection

In Ref. [4] authors define an attribute as "Any instance which is given as input to a machine learning a technique that has fixed values for each of its characteristics". Correlating this concept with databases, attributes are the columns of a DMS, representing the different characteristics of the several instances contained in the database. Instances are often referred records or tuples.

Considering this concept of "attribute", authors of Ref. [5] define the attribute selection process as the determination of a subset of good attributes that will be responsible for generalizing the information contained in the database. The subset of attributes obtained must necessarily have a size equal to or less than the original set of attributes.

Regarding the operation of algorithms for selecting attributes, they usually choose the attributes by an assessment of individual or subsets of attributes. The individual assessment orders the attributes with respect to their relevance. Thus, these methods remove irrelevant attributes, however, do not eliminate redundant attributes. In the case of subset assessments, attributes that are redundant, as well as irrelevant, are removed from the set of attributes [6].

Methods for selecting attributes can be divided into two classes: wrappers [7] and filters [8, 9]. Filters differ from wrappers just on the independence of the algorithm used for the attribute selection [9, 10]. There are a wide diversity of algorithms that perform these tasks. However, only three of these are explained in greater details.

2.2.1. *Attribute selection based on wrapper*

Wrapper methods are commonly used when one wants to select attributes in supervised learning problems. The methodology consists of the input of a attribute set in which the attributes pass through a predetermined search method and evaluation algorithm. Wrapper methods are called this way, because it essentiality wraps up an evaluation algorithm.

The search method is an algorithm that yields different subset of attributes. There are various methods for searching subsets: forward search, where the initial subset starts with a single attribute and then have attributes inserted; backward search: where the initial subset is formed by all attributes and then have attributes removed: bi-directional search: an initial subset will have attributes removed and inserted; exhaustive search: all the possible subsets are obtained.

After the search for a subset of attributes, the wrapper method calculates a performance estimate by inducing the learning algorithm, which will evaluate only those selected attributes.

The search is terminated after a number of performances do not improvement. This parameter is used to escape from local maxima and seek global maxima.

After all the subsets have passed through the same process, the best subsets are selected and will again pass through the learning algorithm. that will extract the most important attributes within the subset, which are evaluated together with a validation set [7]. Finally, with the selected attributes, a classification can be made using the learning algorithm.

In **Figure 3**, it can be seen a brief presentation that generalizes the wrapper methods, to illustrate the procedures discussed above.

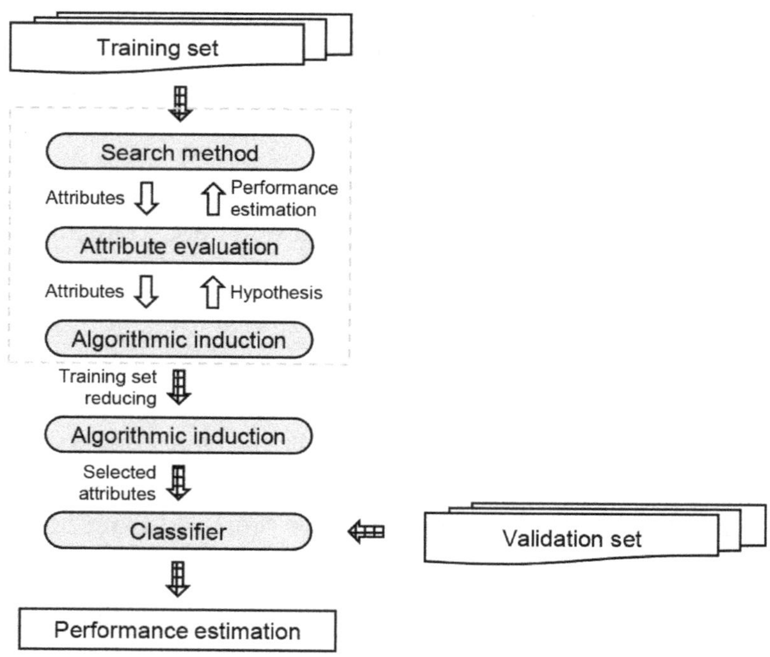

Figure 3. Overall representation of the wrapper attributes selectors.

This algorithm is useful, however, even providing better results than filters; it is slow, i.e., because the learning algorithm is called repeatedly.

2.2.2. Correlation-based feature selection

The correlation-based feature selection (CFS) method has been proposed in Ref. [12], where it can be applied to sets of continuous and discrete data, as shown in Ref. [10]. The method makes use of correlation to estimate the cost of attributes; however, a big difference presented by CFS, when compared to other filters, is that the selection starts by assessing attribute subsets. For each subset, the addition/removal of attributes is then evaluated [11].

The CFS algorithm was proposed according to the following hypothesis: "Good attribute subsets contain attributes highly correlated with the class; however, not greatly correlated to other attributes".

The search for the best subset of attributes ends only when the stopping criterion is satisfied, and this is considered when a predefined number of iterations return the same subset of attributes. Some advantages presented by the CFS algorithm are its rapid implementation, the possibility of application in any set of attributes.

2.2.3. Consistency-based filter

The consistency-based filter (CF) method proposed in Ref. [9] evaluates the attributes of subsets according to their consistency with the classes comprising the data set, and unlike most methods for selecting attributes, it does not use search heuristics, but a probabilistic search algorithm based on Las Vegas algorithm.

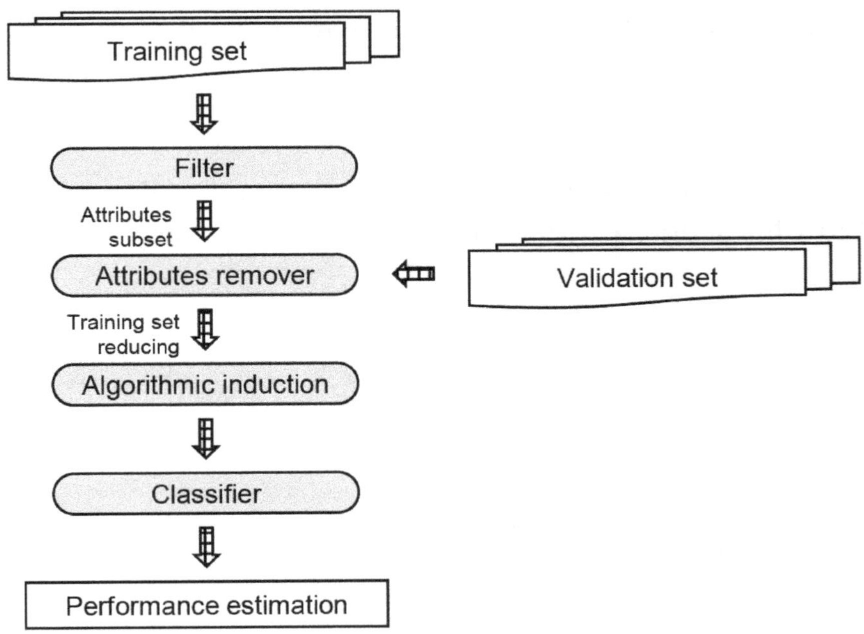

Figure 4. Overall representation of the filter selectors.

From the experimental results obtained by Ref. [9], it can be seen that this method provides a quick response, ensures the location of the optimal attribute subset and is easily implemented.

Las Vegas algorithm makes probabilistic choices to assist the search process. Thus it results quickly in the best attribute subset. This search is performed until a maximum number of attempts is reached. In the end, both the size of the attribute subset and their inconsistency with respect to the class are evaluated. The selected subset is smaller, and consistent since the consistency of an attribute subset is inversely proportional to its inconsistency.

A disadvantage presented by the probabilistic search compared to a heuristic search lies on the computational cost, which is a little higher. However, its biggest advantage is that it does not have the same vulnerability presented by the heuristic search when subjected to data sets with many related attributes.

As an illustration, a block diagram representing the general operation of filters employed in the feature selection task can be seen in **Figure 4**.

2.3. Data Mining

As mentioned before, Data Mining can be considered a step in the KDD [13]. This concept was originated from the need to extract patterns in databases in an attempt to extract valuable information, which could cause companies to become even more competitive within its sector of activity. In **Figure 5**, it can be seen that the Data Mining system is divided into four steps summarized below:

- Step 1 – use of the corporate database or a test database;
- Step 2 – use of techniques for pre-processing data, for example, data normalization (commonly used) and the techniques for selecting attributes of the database;
- Step 3 – from the pre-processed database, Data Mining techniques can then be applied, such as artificial neural networks and fuzzy inference system;
- Step 4 – the response obtained by the Data Mining process, as well as its analysis should be assessed to verify the performance of computational tools employed.

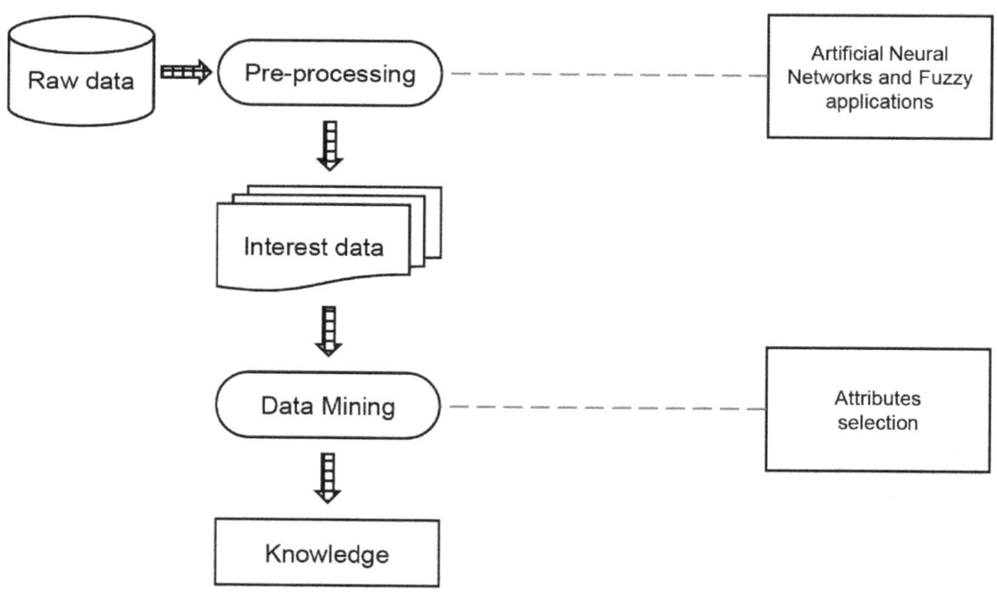

Figure 5. Representation of the KDD process.

Therefore, Data Mining is often used to specify groups of patterns that can be found in databases, without any prior knowledge of which pattern may provide information of interest [13–15].

There are cases where some idea of what kind of information one wants to extract from the database exists. However, the complexity of the extraction task prevents that the analyses are carried out by specialists [14].

The tasks performed by Data Mining are classified as predictive or descriptive [14]. When applying the predictive mining, data representing previous situations of a process are analysed to obtain patterns that may represent the current or future situations of the process.

Since the descriptive mining is used to characterize data contained in the actual database, only current conditions of the process can be determined by this type of mining.

2.4. Data Mining phases

According to Larose [16], Data Mining is considered a cyclical network, because success depends heavily on adjustments made in an ordered set of six phases, totally interrelated, as illustrated in **Figure 6**.

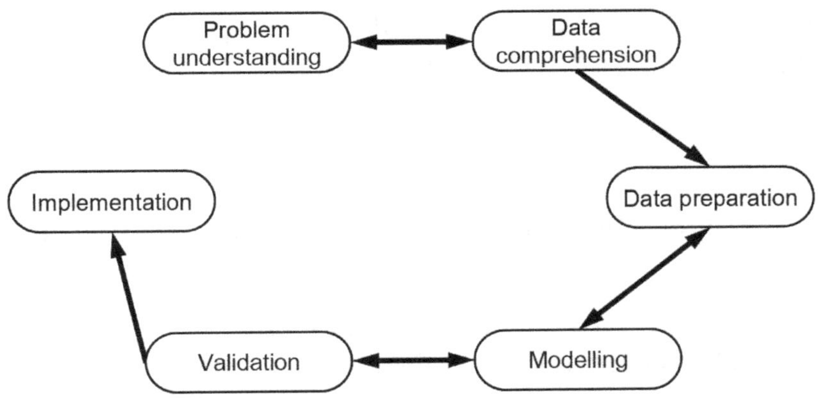

Figure 6. Data mining phases.

It can be seen that the phase of understanding the problem and comprehending the data are bi-directionally connected for information exchange. According to MSDN Microsoft [17], these two phases comprise analysing business requirements, defining the scope of the problem, defining the metrics for evaluating the model and, ultimately, identifying specific data targets for the mining project.

Data can be spread and stored in different formats and may contain inconsistencies, such as missing or incorrect entries. With an appropriate cleaning process and filters, it is possible to fill the empty records. Such actions are carried out during the data preparation phase.

As pointed out in Ref. [16], it is in the modelling phase that algorithms will perform the mining process. The choice of which source to use is dependent on specific application goals.

In the validation phase, models developed in the previous phase will have their performance evaluated as to their performance across a set of data out of the set initially proposed for learning purposes. This stage has a bi-directional connection with the modelling so that fine adjustments can be tuned in mining algorithms.

Finally, after the validating the models, the tools developed in the previous stages can be implemented.

3. Clustering

Currently, tools focused on Data Mining can perform various tasks such as:

- Description provides information that enables the interpretation of the data, therefore, models need to be as transparent as possible to the user. In this case, the quality of information should be the highest possible.
- Estimation provides feature values for the process mapped to the database.
- Prediction allows finding future values of time series from historical information available, even if fitting the function is not possible.
- Classification provides data classification results with respect to previously known classes or categories of data. It is one of the tasks best performed by Data Mining algorithms.
- Association provides relationship rules between attributes, which can have no clearly observable similarities.
- Clustering provides grouping patterns of classes or attributes, which can have clearly observable similarities.

The grouping of standards, records, classes, etc., with similarities, is notably the first approach for managing a mass of unobservable naturalistic data. Clustering is considered as the first task to be performed in order to find common characteristics in databases without knowing the relationship between its entities.

After completing the data assembly process and looking for similarities between instances, further studies involving other tasks associated with Data Mining becomes possible.

Instances spatially distributed without any clearly observable common feature can be clustered by an algorithm that finds elements in common. The algorithms that perform this task are called unsupervised. This is because as this process is performed on data with little or no clear observability, the-the relationship between the entities is unknown.

Most algorithms are based on distance metrics such as the Euclidean distance. These algorithms seek similarities between records through monitoring a parameter that indicates the degree of similarity between the tested standards.

Currently, great efficiency in clustering processes has been achieved with intelligent algorithms, especially with the ones based on Artificial Neural Networks. The Adaptive Resonance Theory (ART) neural network architecture, setup with the recurring features the property to learn new patterns without destroying information learned previously. Therefore, it has a high clustering power and converges very quickly.

4. Database systems

The databases are susceptible to errors caused by a multitude of factors. The data may be incomplete when part of the information is missing, or inconsistent, when the stored information does not match reality [14].

The vast majority of these problems can be exacerbated when the incorrect choice of database, considering the factors listed previously.

Broadly, if it is found that the variety and volume of data currently generated quickly exceed the capabilities of a conventional DMS, the need for tools that can deal with such features is made clear.

We assume that a DMS able to act in Big Data environments must consist of heterogeneous information-storage bases, as exemplified in **Figure 7**.

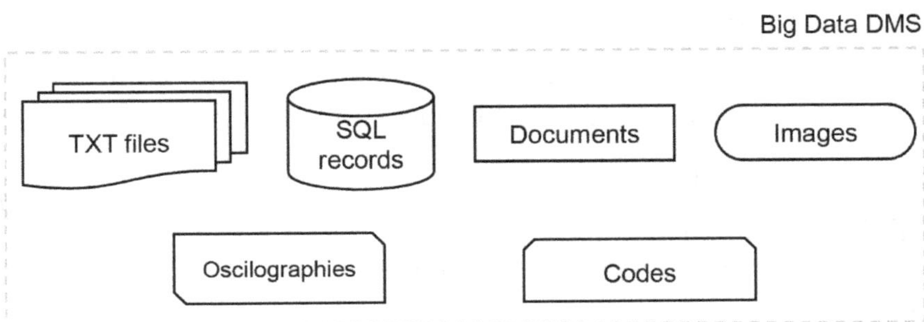

Figure 7. Big Data DMS.

TXT files can contain short reports or even status or device configurations. SQL records can document all kinds of information that can be tabulated, such as customer data, assets, etc. Documents and images have a multitude of applications in companies, which should be properly stored and managed.

Codes of different applications can be stored, either to record update on devices, computers, component firmware, etc. Oscilographies can be responsible not only for protection tasks but also for billing and power quality.

Edgar F. Codd in 1969 proposed a structured model of data storage, which enabled the success of relational models. However, to treat the large volumes, diversity and scale of existing databases, there are new models able to act upon both the structured and unstructured environment.

In this context, it is clear that the use of relational databases for this purpose becomes impractical, giving space to other databases. Among these, we highlight the NoSQL, which is a whole new generation of DMS able to act not only with SQL.

Originally developed to keep up with the web growth in 2009, NoSQL is configured today as a safe and viable option for developers on Big Data as it offers free scheme for development and support, being able to work in environments with massive amounts of data.

We can highlight systems that use the core NoSQL [18]:

- Hadoop/HBase: Written in Java, which allows communication with any protocol.
- MapR, Hortonworks, Cloudera: Professional Hadoop distribution.
- Big Table: Distributed structured data storage system. Used by Google application systems [19].
- Amazon DynamoDB!: Provided by Amazon in its suite of Web Services, being flexible and having consistent latency below 10 ms at any scale [20].
- IBM Informix: Horizontally and vertically scalable relational model from IBM, made of C and contains enhancements to access speed.
- Neo4j: A model based on graphs and represents an evolution of models initially introduced by Facebook and Google.

The functioning of a synthesis Database Manage System (DMS) in Big Data environments does not reflect in the replacement of the traditional models. On the contrary, there should be coexistence between the various forms of information storage that companies already use.

The database mechanism shall function as an integrator of solutions and capable of reliably and quickly supporting high scalable information that is currently generated one of the major works involving the integration of databases can be found in Ref. [21], which contains a study by Harvard Business School.

According to a report conducted in 2010 by the Brazilian Ministry of Mines and Energy, as the related entities, as databases have differently, there is a tendency to adopt models such as XML, for the IEC 61850 defines models for data acquisition based on this language.

5. Data security

5.1. Internal security

Internal security ensures the integrity of the information stored. Every write operation on databases may incur errors, such as conversion errors from A/D converters, syntax error in the symbolic dictionary and physical errors in the recording media.

Ensure that the saved information will be available for future access depends on the technology used but also on the policy of information preservation. It is necessary to ensure user processes efficiently and methodically in data storage or make this process transparent, with a high level of automation. This second option is the one that has been used, and the burden of responsibility transferred to the team that manages the Information Technology.

As the mining process usually requires the integration and transformation of databases, the data internal security depend on several factors, such as [14]:

- Integration schemes: Cohesion of real-world entities;
- Redundancy attributes: Statistical analyses to check correlation between attributes;
- Resolving conflicting data: Differences involving scale, encoding, representation, etc.

Treatment of inconsistencies in the databases can be performed manually, automatically, or with both methods. About 70% of all of the processing time of a DMS is spent searching and correcting errors to ensure the internal security of the information.

5.2. External security

The external security concerns the vulnerability of the information stored in databases. With the introduction of the concept of "Internet of Things", it is assumed that a very large number of IP addresses are available, with the use of IPv6 designation. Indeed the penetration of Smart Grid devices in the daily lives of users is also one of the appeals of these technologies. However, caution is needed with which information will be available to users.

The number of known malicious codes grows day by day, representing a major challenge for the containment team to attack.

The vast majority of databases are currently working with encryption keys, getting the information confined only between devices able to exchange valid keys. Access attempts with invalid keys or even force to obtain valid keys require much computational resource from malicious users. However, this is a real security threat. Mainly because along with the concept of Big Data implemented in the Smart Grid, it is emerging the concept of Smart Cities, where there is the integration of systems and services.

In August 2015, it was held in São Paulo, Brazil, the event "Connected Smart Cities" [22]. One of the topics discussed was the security aspects in the so-called Smart Cities.

Once the idea is the integration of resources, networks, services, etc., such systems need to be separated into physical networks and into different levels of protection.

As communications occur between different entities, there is a targeting component searches for open protocols are adopted, such as based on the TCP/IP family.

NIST considers research in the area of information security still very incipient for the effective implementation of Smart Cities.

The challenges are reflected in the form of encryption key management systems with volumetric applied to Big Data, with also exponential scalability. As the integration will arrive at various levels not previously experienced by companies, secure protection alternatives eventually must face the balance between performance and redundancy.

Systems with great safety tend to be slow and thus provide numerous other types of problems such as overloading of communication channels, increased error rate, idle time processing system tasks. We can list a few critical factors for the protection of these databases:

- Segmentation of networks: An increased presence of schemes that correctly isolates network is fundamental for service integration of services.

- Redundancy: The information, which will travel between the entities, requires authentication mechanisms with low vulnerability to fraud.
- Enforcement of countermeasures: The use of security measures needs to be in accordance with all kinds of communications that will be employed by the company, such as firewalls, authentication certificates for VPN networks and security keys.
- Investment in human assets: Constant training will be needed so that those responsible for information security are always able to work in the services integration environment of Smart Cities.

Author details

Danilo Hernane Spatti[1*] and Luisa Helena Bartocci Liboni[2]

*Address all correspondence to: danilospatti@utfpr.edu.brs

1 Federal Technological University of Paraná, Curitiba, Brazil

2 Federal Institute of Education, Science Technology of São Paulo, University of São Paulo, São Paulo, Brazil

References

[1] Mayer-Schönberger, Viktor; Cukier, Kenneth. Big data: A revolution that will transform how we live, work, and think. Houghton Mifflin Harcourt, 2013.

[2] Appel, A. P. (2010). Pre processing and data mining methods for great amount of data and complex networks. PHD Thesis, University of São Paulo.

[3] Consularo, L.A. (2000). Data mining techniques for image analysis. PHD Thesis, University of Sao Paulo.

[4] Witten, I., Frank, E., Data Mining – Practical Machine Learning Tools. Morgan Kaufmann, 2005.

[5] Liu, H., Motoda, H., Feature Selection for Knowledge Discovery and Data Mining. Kluwer Academic Publishers, 1998.

[6] Lee, H.D. (2005). Major attribute selection for knowledge database extraction. PHD Thesis, University of Sao Paulo.

[7] Kohavi, R., John, G.H., "Wrappers for feature subset selection". Artificial Intelligence, vol. 97, pp. 273–324, 1997.

[8] Almuallim, H., Dietterich, T.G., "Learning with many irrelevant features". Proc. of the 9th National Conference on Artificial Intelligence, pp. 547–552, 1991.

[9] Liu, H., Setiono, R., "A probabilistic approach to feature selection: a filter solution". Proc. of the 13th International Conference on Machine Learning, pp. 319–327, 1996.

[10] Hall, M.A., "Correlation-based feature selection for discrete and numeric class machine learning". Proc. of the 17th International Conference on Machine Learning (ICML), pp. 359–366, 2000.

[11] Hall, M.A., Holmes, G., "Benchmarking attribute selection techniques for discrete class data mining". IEEE Transactions on Knowledge and Data Engineering, vol. 15, no. 3, pp. 1–16, 2003.

[12] Hall, M.A., Correlation-based Feature Selection for Machine Learning. Ph.D. Thesis, The University of Waikato, Hamilton, New Zealand, 1999.

[13] Fayyad, U.M., Piatetsky-Shapiro, G., Smyth, P., Uthurusamy, R., Advances in Knowledge Discovery and Data Mining. AAAI Press, 1996

[14] Han, J., Kamber, M., Data Mining: Concepts and Techniques. Morgan Kaufmann, 2001.

[15] Oliveira, S.R. (2004). From data to knowledge: Evolution and challenges. Associated Professor Thesis, University of Sao Paulo.

[16] Larose, D.T., Discovering Knowledge in Data: An Introduction to Data Mining. Wiley, 2005.

[17] MSDN Microsoft, "Data Mining Concepts". https://msdn.microsoft.com/pt-br/library/ms174949(v=SQL.120).aspx

[18] List of NoSQL databases. Accessed in September 2016, available in http://nosql-database.org

[19] Chang, F., Dean, J., Ghemawat, S., Hsieh, W.C., Wallach, D.A., Burrows, M., Chandra, T., Fikes, A., Gruber, R.E., Bigtable: A Distributed Storage System for Structured Data. Google Inc, 2006.

[20] Amazon DynamoDB. Accessed in September 2016, available in http://aws.amazon.com/pt/dynamodb/

[21] Davenport, T.H., Harris, J.G., Competing on Analytics: The New Science of Winning. Harvard Business Press, 2007.

[22] Connected Smart Cities. Accessed in September 2016, available in http://www.connectedsmartcities.com.br

9

The Role of Communication Technologies in Building Future Smart Cities

Abdelfatteh Haidine, Sanae El Hassani, Abdelhak Aqqal and Asmaa El Hannani

Abstract

The world population is continuously growing and reached a significant evolution of the society, where the number of people living in cities surpassed the number of people in rural areas. This puts national and local governments under pressure because the limited resources, such as water, electricity, and transports, must thus be optimized to cover the needs of the citizens. Therefore, different tools, from sensors to processes, service, and artificial intelligence, are used to coordinate the usage of infrastructures and assets of the cities to build the so called *smart cities*. Different definitions and theoretical models of smart cities are given in literature. However, smart city can usually be modelled by a layered architecture, where communication and networking layer plays a central role. In fact, smart city applications lay on collecting field data from different infrastructures and assets, processing these data, taking some intelligent control actions, and sharing information in a secure way. Thus, a two way reliable communications layer is the basis of smart cities. This chapter introduces the basic concepts of this field and focuses on the role of communication technologies in smart cities. Potential technologies for smart cities are discussed, especially the recent wireless technologies adapted to smart city requirements.

Keywords: ICT, smart cities layered model, quality of service, communication requirements, low-power wireless area network, power line communication, machine-to-machine communication, Internet of things

1. Introduction

The world population has grown exponentially at an average rate of 1.2% per year in the last 50 years. In 2007, for the first time ever, the number of people living in cities surpassed the number of people living in rural areas. Furthermore, it is estimated that the proportion will exceed 70% by 2050. According to the UN World Economic and Social Survey for 2013, Africa, Asia, and other developing regions will be housing an estimate of 80% of the world's urban population in the coming years [1]. While urbanization has advantages, it also brings challenges. Rapid urbanization adds pressure to the resource base and increases demand for energy, water, and sanitation, as well as for public services, education, and health care. Consequently, social, economic, and environmental issues have become tightly interconnected. Cities greatly contribute to environmental degradation on local, regional, and global scales. Studies have demonstrated that they are accountable for 70% of global greenhouse gas emissions as well as 60–80% of global energy consumption [2]. Thus, the challenge for the humanity is that the governments and any decision-maker is to find solutions and means to make roadmaps into execution to answer the question: how can cities be made smart and sustainable under such underlying conditions?

The answer lies in making cities "smarter" by efficient management of resources and infrastructure, greener environment, and smart governance resulting in a better quality of living of its citizens. This can be enabled by the effective use of information and communication technologies (ICTs) tools, which have the ability to provide eco-friendly and economically viable solutions for cities. Potential advancements could be made, among others, through efficient water management based on real-time information exchange, public transport systems organized through information from satellites, air quality, and electromagnetic field monitoring. This is where the concept of smart sustainable city comes into play [3].

Decades ago, cities started to use ICT to provide better services and quality of life to citizens. The evolution of the cities begun from "wired Cities," going through "virtual cities," to "intelligent cities," "information cities," "digital cities," "sustainable green cities," and then "smart cities." Currently, specialists are talking about smart sustainable cities (SSCs). Historical details of the evolution of cities from classic to smart can be found in [4].

There are several definitions of smart cities depending on the research field. In this chapter, only definitions from bodies and organization related to ICT are taken into consideration, namely the ITU and IEEE. The ITU—Telecommunication Standardization Sector (ITU-T), through its ITU-T's Focus Group on SSC, adopts the following definition of a smart sustainable city as "*A Smart Sustainable City efficiency of urban operation and services, and competitiveness, while ensuring that it meets the needs of present and future generations with respect to economic, social and environmental aspects*" [3]. The IEEE, in the context of Internet computing, states that a smart city can effectively process networked information to improve outcomes on any aspect of city operations. Such operations cover a very broad domain: surfacing information to authorities, businesses and citizens, and optimizing energy and water production or consumption, traffic management, public safety, and emergency response [5]. To address smart cities' multifaceted and cross-domain challenges, the Internet plays a fundamental role in communication, information sharing and processing, data transfer and analysis, and distributed computing. The rise of the Internet of things (IoT) and the large-scale adoption of Web technologies in urban environments have proven that Internet-based solutions can successfully address societal challenges.

For the elaboration of roadmaps to design and deploy smart cities, a layered architecture is used to model the smart city with its different application fields and infrastructures. Communication and networking layer as the middle of this layered architecture plays a very important role in building smart cities. This is why we focus on this central layer by discussing in this chapter the related requirements and communication technologies. The discussion is illustrated by the layered model of smart grid, as it is one the most critical national infrastructures (CNIs), since all industries, administration, and citizen's life are strongly depending on it. The chapter is closed by an overview on the most recent advances in wireless communications, which adapted to build the backbone of the smart cities.

2. Applications of smart cities and the enabling ICT

2.1. Overview of smart cities applications and features

Nowadays, there are more sources of innovation available in the world than ever before. By using the accumulated progress and improvements in ICT during the last decades, the smart city will provide relevant scenarios of livability for citizens with the necessary quality and simplicity in urban areas. A further purpose of the smart cities could be also to drive economic growth and area-based development in order to build a clean and sustainable environment, enhance incomes for all, support disadvantaged persons, or to make the governance of the city more transparent. If so, smart cities' paradigms should extend and/or radically transform our traditional infrastructure by bringing together innovation, reliability, mobility across the city, and easiness of daily services in the urban spaces. As depicted in **Table 1**, smart city applications are then a mesh of technologies services and a combination of many vertical smart solutions deployment in different locations and over many domains, such as health care, logistics, commercial and industrial networking, security and surveillance, transportation and mobility, education.

2.2. Smart cities infrastructures and role of ICT

Setting up a smart city is more than improving the old system with technology by simply adding sensors, remote supervision, and control to essential city services. It should be a complete shift of a paradigm in daily life when using new technologies, especially new ICT leading to smart outcomes. If that is the case, some major questions must be asked about the adoption of any transformation strategy to retrofit smart city ideas into the existing city: What are particular opportunities and threats on the existing infrastructure and services? How to redesign and integrate ICT in a smart way? How to manage the risk of growing volumes of all the gathered information? How could that transform and automate information usage in business processes? How to use ICT to enhance the quality of life and to better engage residents to such integrated vision for the improvement?

Smart energy	Smart buildings	Smart transportation	Smart water
- Smart metering - Demand response and demand-side management - Distribution automation - Distributed generation - Integration of renewables and decentralised energy - Network monitoring and control	- Light control - Heating control - Energy efficiency - Local energy generation - Security, occupancy control - Synergies between energy efficiency, comfort and safety and security - Building as a network: Integration of multiple technologies (HVAC, lighting, plug loads, fire, safety, mobility, renewable, storage, materials, IAQ, etc.) - Software: Efficiency, automation and control, analytics and big data management	- Vehicle-to-everything (V2X) - Driver behaviour management - Mobile applications based on open data - Traffic and fleet monitoring and control - Services for drivers and passengers based on real- or near-real-time information - Integrated public transportation	- Pressure management - Remote control and predictive maintenance - Integrated platforms for water management - Smart metering - water conservation and efficiency
Smart waste	**Smart physical Safety/security**	**Smart health care**	**Smart education**
- Waste Management - Wastewater treatment - City cleaning - Sorting of waste - Waste tracking	- Video surveillance and video analytics - Seamless communication during natural and man-made disasters	- Adequate sanitation - Disease control and mitigation - Smart hospitals - Real-time health care including analytics - Home and remote health care including monitoring - Electronic records management	- Flexible learning in an interactive learning environment - Accessing world class digital content online using collaborative technologies - Massive open online course (MOOC)

Table 1. Main applications related to infrastructures of smart cities.

Certainly, answers will be different in every case as there is no unique and generic approach for a simple step-by-step procedure in building smart cities using ICT. As a result, we typically start by the modeling of the smart cities. Since many possible solutions exist, there are discussions about the main pillars building the city, after that the smartness may be checked in each pillar of the city. Without loss of generality, the main common four pillars building the city in the literature are economy, governance, environment, and society, according to [3]. These are reflected via three overarching dimensions of a city: environment and sustainability, city level services, and quality of life. Each of these dimensions has multiple characterizing attributes, sometimes overlapping. Sustainability and environment are critical to the urban landscape since cities represent 75% of energy consumption and 80% of CO_2 emissions on a global basis. The primary attributes in this dimension include infrastructure and governance, energy and climate change, pollution, waste, social, economic, and health aspects. As for city level services, the key attributes include technology and infrastructure (e.g., transportation, buildings, health care), sustainability (e.g., water, air, waste), governance (e.g., organization, administration, and leadership) and economy (e.g., financial, human capital, economic strength). The final dimension is the quality of life of the citizens. This reflects how the inhabitants of a city perceive their own sense of well-being and the fact that they are constantly striving to better themselves—for example, in terms of wealth, health, and education. All of the above need to be balanced for a successful smart sustainable city [3].

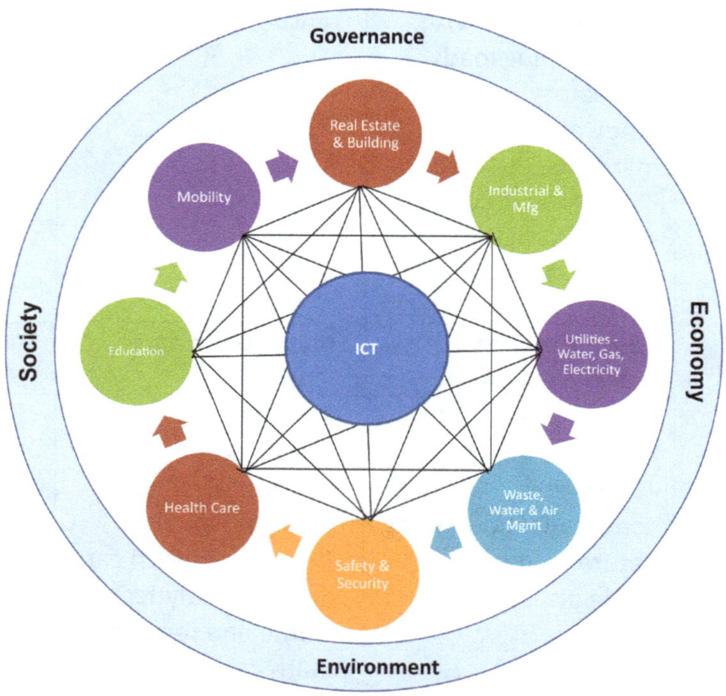

Figure 1. Core field components of sustainable smart cities and position of ICTs.

Infrastructure is a pivotal aspect of a smart sustainable city. Traditionally, there have been two types of infrastructure: physical (e.g., buildings, roads, transportation, and power plants) and digital (information technology (IT) and communications infrastructure). There is also a service infrastructure, providing services run on top of the physical infrastructure (e.g., education, health care, e-government). The digital infrastructure provides the glue enabling the smart sustainable city to operate efficiently in an optimal way.

ICT has a crucial role in SSC since it acts as the platform to collect and aggregate information and data from the field to help enabling an improved understanding on how the city is functioning in terms of resources consumption, services, and lifestyles, as illustrated in **Figure 1**. ICT also enable the following functions, which are keys to achieving the goals and maximizing the performance of smart city [3]:

- ICT-enabled information and knowledge sharing: Traditionally, due to inefficiency on information sharing, a city may not be ready to solve a problem even if it is well equipped to respond. With immediate and accurate information, cities can gain an insight on the problem and take actions before it escalates.
- ICT-enabled forecasts: Preparing for stressors like natural disasters requires a considerable amount of data dedicated to study patterns, identify trends, recognize risk areas, and predict potential problems. ICT provides and manages this information more efficiently, so that the city can improve its preparedness and response capability.
- ICT-enabled integration: Access to timely and relevant information (e.g., ICT-based early warning systems) need to be ensured in order to better understand the city's vulnerabilities and strengths.

3. The smart city architecture: a multilevel model and a use case

3.1. Layered architecture of smart cities model

The realization of smart cities involves different components and players: sensors vendors, equipment vendors, communication providers, services providers, business innovation. Users of smart cities belong to different domains, such as industry, utilities, transport/logistic, health care. These different components can be organized in a layered architecture, as illustrated in **Figure 2**, which shows the importance of ICT in building smart cities by connecting the applications/services and the city infrastructures, such as electricity grid, roads, parking. Much of the current debate about smart cities concerns the data layer, and some of the most important smart city work currently being explored in trials, demonstrations, and projects is to solve issues of data security, privacy, trust, sharing, standardization, APIs, and commercialization. There are opportunities for multiple players here to help cities to understand what data they have and how they can accomplish collection, enrichment, analysis, and use of their data effectively in practice. However, while recognizing the significance of data, the main focus of this chapter is the communication networks layer.

Basically, communication networks are hardware and software entities that could support and facilitate communication of any form (video, text, data, or any other digital objects) from a local area network (LAN) to wide area network (WAN).

Thus, why is the focus on communication networks so important in the case of a smart city? What, exactly is ICT supposed to do in this case? In fact, the layer of communication networks is the midpoint, the main vehicle of information and the core framework to model how applications and services can communicate over one or multiple networks and to set how data communication should take place when using applications of smart city. It defines and describes the way in which interrelated entities communicate more effectively between each other; systems arranged and connected to emerge smartness in use and provide higher quality services. In other words, effective deployment of ICT, if possible, may be a good factor of making easy modular engineering in building smart cities and understanding relationships of all the information we are gathering. However, it is important to notice that we cannot simply dump these ICT opportunities and city cases into the hands of the network operators and expect them to rush off to develop a system that really satisfies all the needs and technical specifications. We will probably need to be a lot more definitive about what we want the communication networks to do. Of course, ICT do not define itself a solution, but it provides the first perspective on what any viable solution would need to accomplish a move to a smart application using the latest tools and technology.

Therefore, an attempt should be made to deeply understand the role of such communication networks in the support of smart city applications. As sometimes the concepts of smart cities are too large with regard to communication networks scenarios, we should keep things simple. We might find it beneficial to make our explanation more concrete through an illustrative use case as a part of the overall process. The smart grid is examined in the following as a use case.

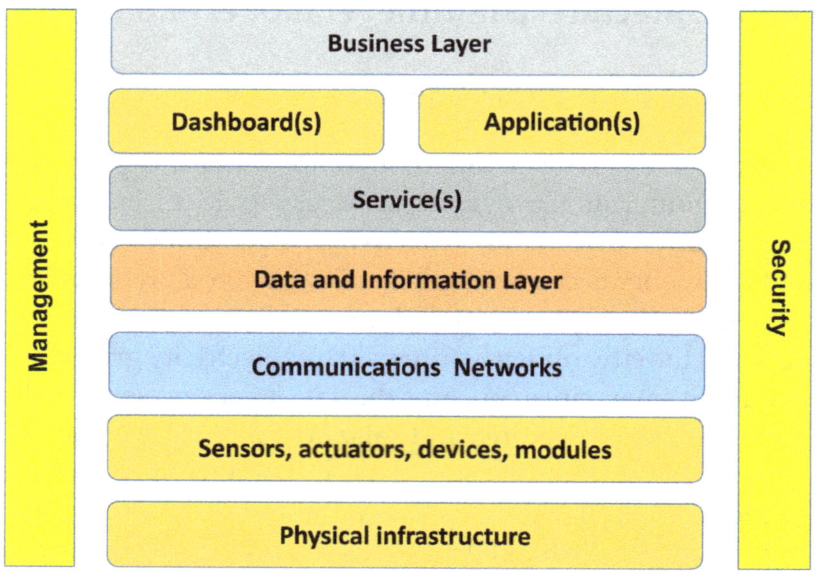

Figure 2. Functional layers and of smart cities.

3.2. Smart grid as an illustrative use case of smart cities

The energy infrastructure is considered to be the single most important part in any city. Therefore, electricity grids are set as *critical national infrastructure* (CNI). If unavailable for a significant enough period of time, all other and also critical organizations/functions are affected, such as security, police, telecoms. Smart grid includes intelligence in control and monitoring of the electricity and water infrastructure in all levels, from the generation plant to the consumer premises through transmission and distribution. Smart grid is defined by IEEE as *"An automated, widely distributed energy delivery network characterized by a two-way flow of electricity and information, capable of monitoring and responding to changes in everything from power plants to customer preferences to individual appliances."* A smart grid covers three major classes of functions: (a) It modernizes power systems through self-healing designs, automation, remote monitoring and control, and establishment of microgrids; (b) it informs and educates consumers about their energy usage, costs, and alternative options, to enable them to make decisions autonomously about how and when to use electricity and fuels; and (c) it provides efficiently safe, secure, and reliable integration of distributed and renewable energy resources, where consumers are also participating actively in the energy market. A global, but not complete, picture of smart grid applications is illustrated in **Figure 3**. All these add up to an energy infrastructure that is more reliable, more sustainable, and more resilient. Thus, a smart grid sits at the heart of the smart city, which cannot fully exist without it [6]. Therefore, the example from the practice given in this chapter are from smart grid field, especially from smart metering that is the corner stone of smart grid.

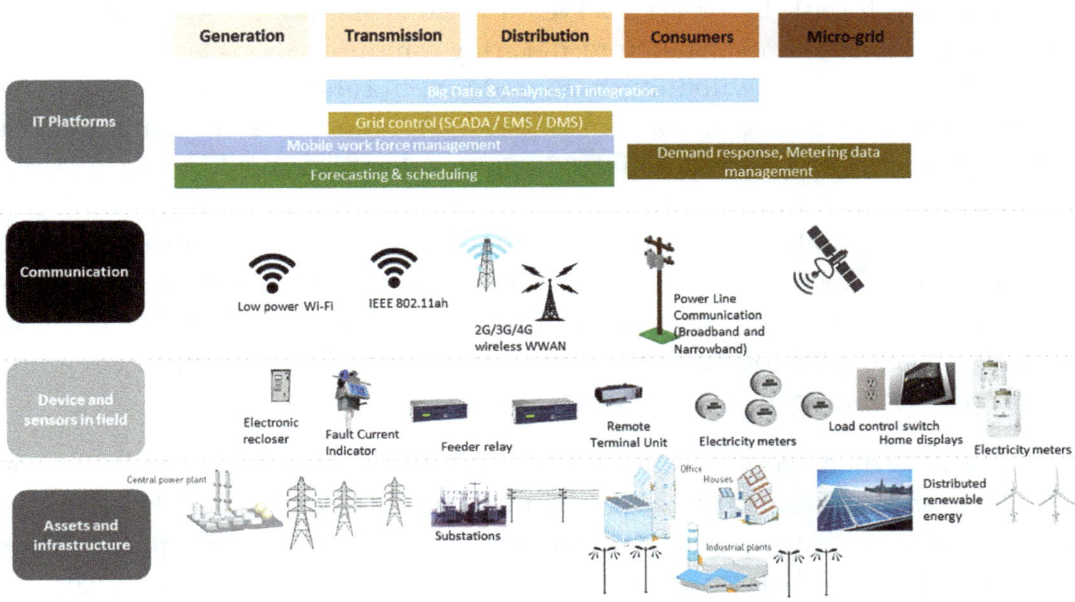

Figure 3. Different levels of smart grid model.

4. A roadmap of using ICT towards smart cities

4.1. ICT requirements in smart cities

As stated before, the ICT infrastructure plays a critical role as the central nerve of the smart city, by connecting and coordinating all the different interactions between the pillars and the infrastructure elements. It is an essential ingredient, since it acts as the "glue" that integrates all the other elements of the smart city in the form of a foundational platform that is present in all interactions of smart city components, either in form of human-to-human, human-to-machine, or machine-to-machine communication. Therefore, the ICT platform must respect a set of requirements in order to accomplish the desired tasks without becoming a cause of malfunctions and constraints of daily activities in the city. The most important requirements are the following:

- Cost-effectiveness: Sensors or metering points, like smart metering communications technologies that will be deployed on a massive scale must be designed to minimize those costs and provide excellent value. The costs-effectiveness of the communication infrastructure should cover all three aspects: capital expenditures (CAPEX) and operations expenditures (OPEX), such as low energy consumption and long battery lifetime, minimal infrastructure, low HW/SW equipment costs and low installation costs and low maintenance.

- High reliability and long lifetime: Communications equipment and modules could be placed in different environment depending on the application. By taking the example of smart grid, smart meters and the communications technologies supporting them should be able to stay online for 10, 15, or even 20 years without requiring the components to be replaced or directly maintained. Therefore, metering communications technologies must be highly reliable and designed with industrial-grade specifications to meet extreme environmental conditions of shock, corrosion, temperature, vibration, humidity, and interferences with other equipment and machines in the deployment environment. A negative example of low-reliable communication solution is power line communications (PLCs), which performances are fluctuating according to activities and machines connected to the electricity installations. Generally, business cases for utility solutions are often based with a 20 years life span for meters, while telecommunication solutions are moving every 2 years. So utilities have a challenge to operate, maintain, and manage the communication infrastructure over a so long time span.

- Network security: When two-way command and control systems are embedded into energy management systems, several security threats must be addressed. Generally, security needs to be guaranteed: confidentiality of data, which only allows authorized entities to access the data; integrity of data, which assures that data are not manipulated; authenticity of data, which guarantees that the data are sent by a dedicated entity; and availability of data, guaranteeing that data are available when needed.

- Interoperability and open standards/SW: Communication systems should be based on standard metering protocols to assure interoperability with changing energy supplier equipment and/or consumer equipment over the life of the sensing end-device. True efficiency is the result of interoperability that allows data to flow freely across various technologies within city functions such as lighting, transportation, and infrastructure—enabling intelligent communication, while maintaining security protocols.

- Quality of service (QoS): The term of QoS covers several criteria used to evaluate and select the communication technologies. The most important of them are as follows: Throughput or bit rate, network capacity in terms of number of connected devices at the same time, delay, jitter.

4.2. Process to select optimal communication technologies

Even with a good communication networks in a city, as when it has built its own fiber network or when it counts multiple mobile network operators (MNOs), or puts in place dense networks of sensors—connectivity is still seen by cities as a challenge. This challenge has different aspects, such as the deployment costs, the costs for operation and maintenance, the diversity of required services and generated traffic. For instance, the last mile connection between sensors and the core communications infrastructure can be made either on an application-specific basis, or in a way that supports multiple applications. Therefore, multiple networks and technologies are required to connect the smart city components as it comprises a collection of Internet of things (IoT) applications, having each their own specific communication and data requirements. For example, the mobile cellular network offers currently over 90% of coverage in cities, but a cellular network is highly unlikely to be able to deliver appropriate connectivity for every smart city application, even if it is able to satisfy many requirements. Examples of challenges for the traditional cellular networks (GSM/GPRS/UMTS/LTE) are the absence of service level agreement (SLA), deep indoor coverage and signal building penetration, high battery power consumption. Despite the emerging narrowband IoT (NB-IoT) networks and partnerships with competing IoT low-power wireless area networks (LPWANs) providers, for the foreseeable future, a smart city will use multiple network technologies.

Different factors must be taken into consideration during the selection of the adequate communications technology for the smart city applications. Therefore, the selection process is complex and difficult to model. However, the main influencing factors can be categorized in three principal axes as illustrated in **Figure 4**:

- **Regulatory and standards**: The international standardization bodies define the technological standards for the design and manufacturing of the technologies equipment and devices. However, the national regulatory has the full control on the use of the infrastructure and the standards allowed to be used in its territory. For example, when designing a wireless networks, only a very limited spectrum of slots are available. Furthermore, the use of available frequencies will be allowed only after purchasing a license from the regulatory authorities. A more concrete example is the 2G/GSM where we can use according to the standard a very large spectrum range going from low frequencies around 400 MHz until 1.9 GHz. However, in each country only very limited frequencies are used, because each country has its own spectrum occupancy and allocation policy as well as the historic.

- **Technical criteria**: This covers all QoS parameters discussed earlier, going from throughput to jitter. The choice of the technology to be used depends on the end application and its requirements in terms of offered throughput and QoS.

- **Strategic decisions**: This helps to check whether the resulting investment will enforce the position of the network owner/operator in the economy landscape of the country, or revenues will be generated to covers the invested money (known as return-on-investment).

The communication network to be built should fulfill the requirement for the applications with minimum costs, considering CAPEX as well as OPEX.

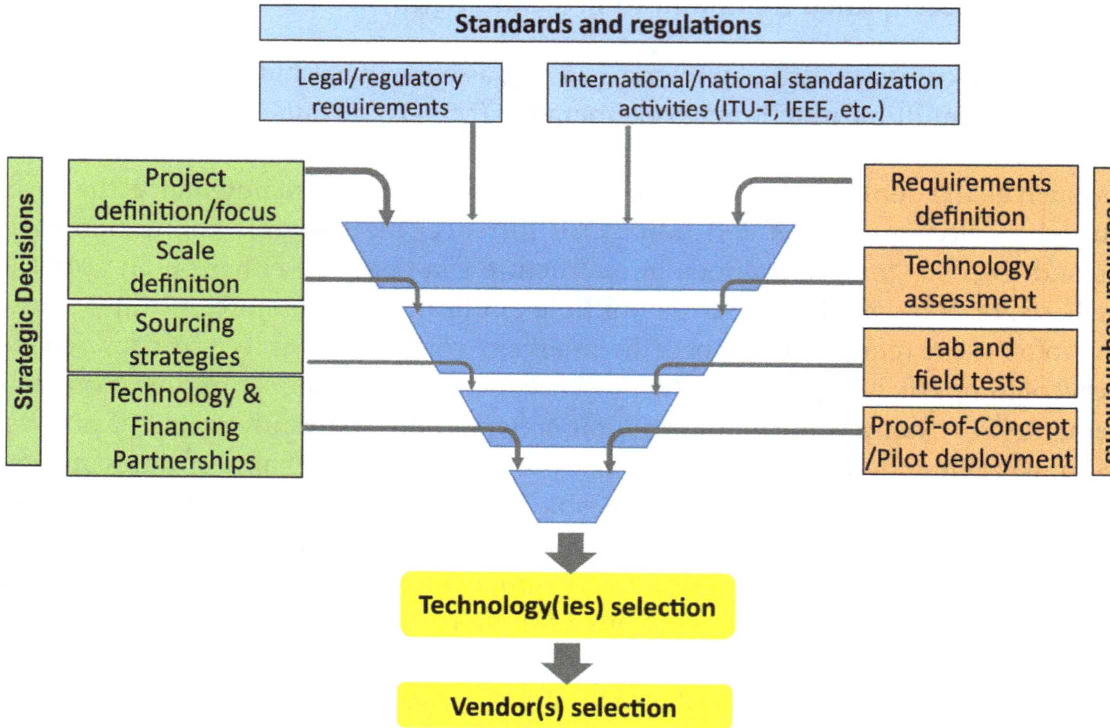

Figure 4. Selection process of communication technologies for smart city applications.

In the next sections, the most adequate technology candidates will be discussed by underlining their advantages, drawbacks, and the smart city applications to which they match optimally.

5. ICT enablers for smart cities

5.1. Machine-to-machine (M2M) and Internet of things (IoT)

M2M communication is sometimes confused with IoT as it is similarly meant to connect sensors and other devices to ICT systems through wired or wireless networks. However, IoT stands for a whole paradigm including a set of technologies, systems, and design principles associated with the emergence of general Internet-connected things, as well as a broader "Internet" network, based on a specific physical environment. IoT ecosystem is expected to reach a structure similar to today's Internet, where content can also include things and real world, with the M2M acting as an enabler. IoT is expected thus, in addition to connecting people, media, and content, to connect all real-world assets, equipped with a variable intelligence enabling information exchange, interaction with people, enterprises business process support, and creativity. Therefore, IoT is an extension to the existing Internet, with an automatic data collection, control, and supervision of physical infrastructure, through a remote monitoring and control, as illustrated in **Figure 5**. A detailed analysis of differentiating characteristics between M2M and IoT is developed in [7]. On the long term, the adoption and deployment of IoT solutions in a wide scale will have to go through a deep understanding of the targeted systems, the economies of scale, and interoperability methods, in addition to key business drivers and governance structures across value chains.

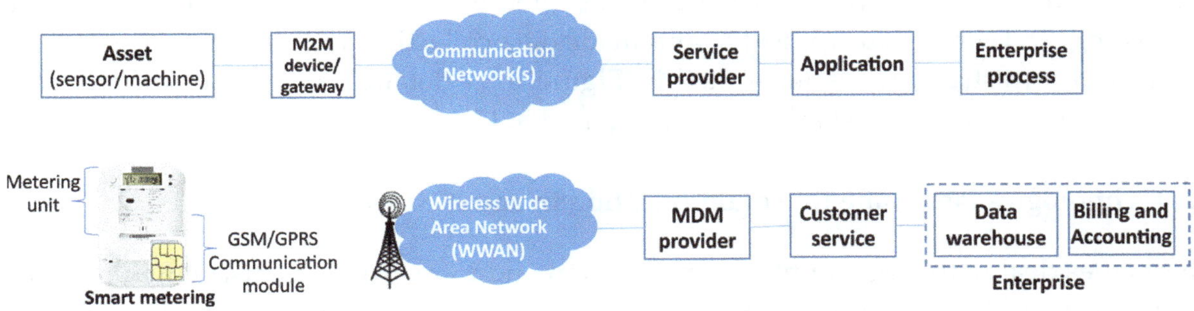

Figure 5. Generic M2M system (top) and corresponding system in smart metering (bottom).

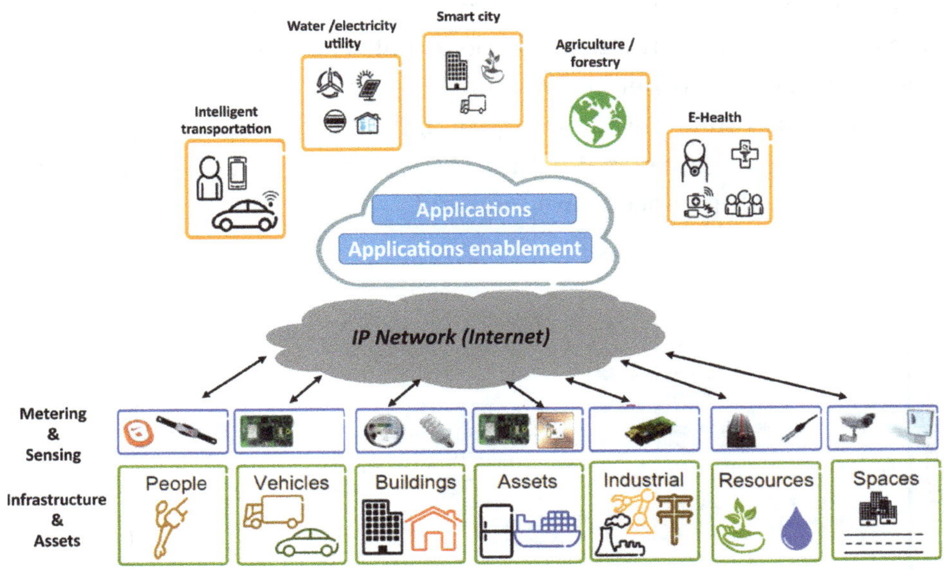

Figure 6. Overview on IoT architecture including different application areas.

The Internet of things (IoT) is a recent paradigm and widely used term for a set of technologies, systems, and design principles associated with the emerging wave of Internet-connected things that are based on the physical environment. In many respects, it can initially look the same as M2M communication, connecting sensors and other devices to ICT systems via wired or wireless networks. In contrast to M2M, however, IoT also refers to the connection of such systems and sensors trough Internet-protocol (IP) networks. This will guarantee a high interoperability and connectivity of devices from different manufacturers. This will increase the mass production and deployment, and hence the price reduction in innovative products, by an IoT ecosystem that will emerge not dissimilar to today's Internet, allowing things and real-world objects to connect, communicate, and interact with one another in the same way humans do via the web today.

In the future, the Internet will not be only about people, media, and content, but it will also include all real-world assets as intelligent creatures exchanging information, interacting with people, supporting business processes of enterprises, and creating. The IoT is not a new Internet, and it is an extension to the existing Internet, where the technology, the remote monitoring, and control are applied to automatic data collection and control and supervision of physical infrastructure, as illustrated in **Figure 6**. A detailed analysis of differentiating characteristics between M2M and IoT is developed in [7].

5.2. Technologies landscape for communication in smart cities

Smart cities deal with large number of population and infrastructures distributed unequally in the city area. This will result in a large number of applications end-points. Therefore, it is normal that different communication technologies will be in use. The common technologies that are currently widely used and will have an important impact in the ICT landscape are presented in **Figure 7**. They are classified into indoor solutions, where sender and receiver are in a relatively small area like room/home/building, and outdoor solutions where the communicating systems are far from each other. It is clear that wireless communications is the mostly used technology, and even Internet is nowadays mostly accessed either via Wi-Fi or 3G/4G. The wireless connectivity offers the advantage of mobility and ease to deploy. However, wired communications will also be present in the communications landscape for smart cities, especially where the very high speeds are required.

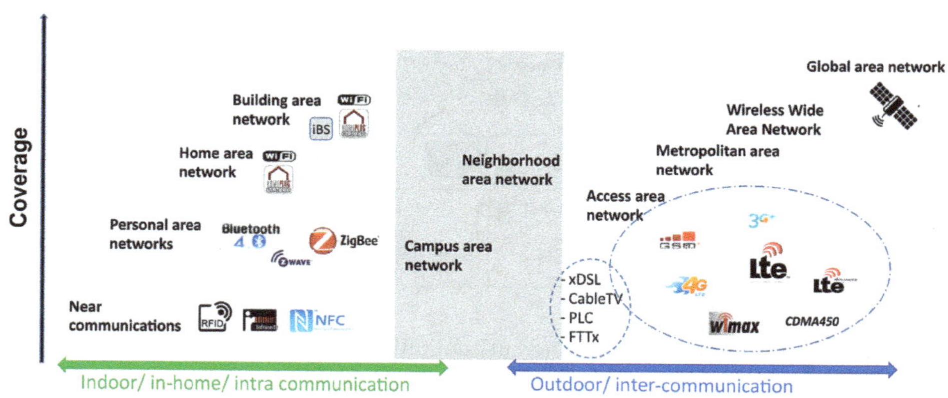

Figure 7. Main technologies to play a role in smart cities communications.

5.3. Wired technologies

As stated above, wired communication is a part of the smart city communication platform. Generally, it offers a high bit rate, immunity against interferences, performance stability in terms of achieved bit rates and Bit Error Rate (BER), unlike wireless communications that suffer from wireless channel impairments and the physical phenomena weakening the electromagnetic wave during the radio-wave propagation, such as reflexions, refractions, path loss, and fading. However, its main drawback is the necessary civil engineering works to lay cables, which generates higher investments and longer deployment time. This was and will be the main obstacle for deploying fiber-to-the-home (FTTH). Therefore, there is a big interest to make reuse of existing cables in the field, like in digital subscriber line (DSL), cable TV (CATV), and power line communications (PLCs). In the normal telecom services for voice and Internet, the DSL solutions have the largest part of the market for the access area networks. However, with the apparition of smart cities and some of its special applications, new technologies have emerged to fulfill some special requirements. For example, PLC is widely used in Europe for the automated metering infrastructure (AMI) that builds the core of the smart grid, because utilities want to use their own infrastructure, in this case the low-voltage power cables, to build their communication platform. By avoiding infrastructure from third parties, utilities aim to have a full control on the communication infrastructure, or at least a part of it. For this purpose, new standards have been developed and new spectrum has been explored, like the standards meters and more deployed for more than 30 million of smart meters by ENEL in Italy [8], which reference architecture is presented in **Figure 8**, the standard PRIME developed by the Spanish utility IBERDROLA [9] and the G3-PLC developed in France by ERDF [10]. Cable TV is also used for smart metering, but the communication between the meter and the cable modem occurs over a wireless local area network (WLAN) through IEEE802.11 technology (Wi-Fi).

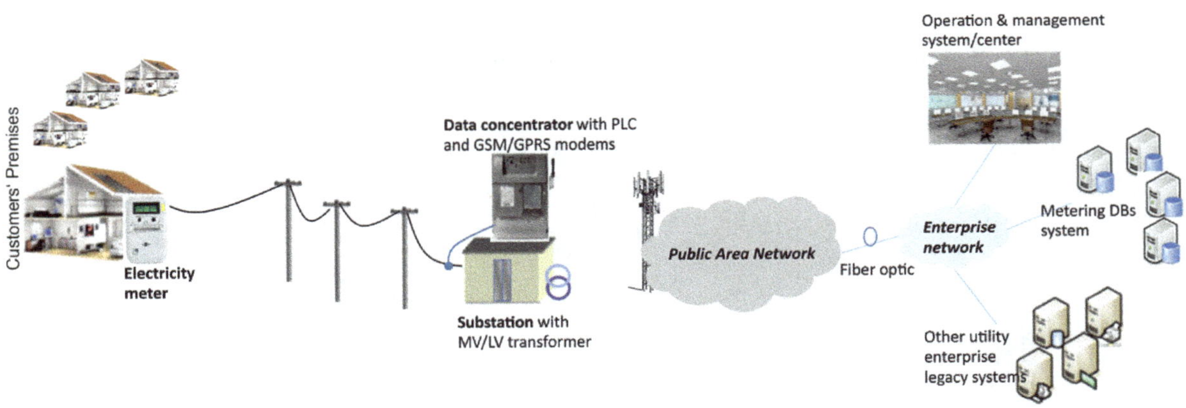

Figure 8. Reference architecture smart metering using PLC.

5.4. Wireless technologies

5.4.1. Overview

Wireless technologies are living prospectus era, where new standards and releases are developed and issued each year. This new era has been intensified since the apparition of the first standard (Release 8) of long-term evolution (LTE), which was handled as fourth mobile generation (4G), as illustrated in **Figure 9**. Since then, the bit rates of mobile communications did not cease in evolving drastically, so that currently 4G are reaching easily 600 Mbps in cities, and even 1 Gbps has been measured in field trials of 5G (**Figure 10**). These explosive throughputs are becoming possible by adopting advanced techniques, like advanced multi-carrier modulations, multi-input-multi-output (MIMO), adaptive coding and modulation (ACM) [11].

In spite of these advantages, some critical points are still challenging the use of commercial mobile services in some smart cities applications. By taking the example of smart grid, the utilities are still seeing the following critical challenges:

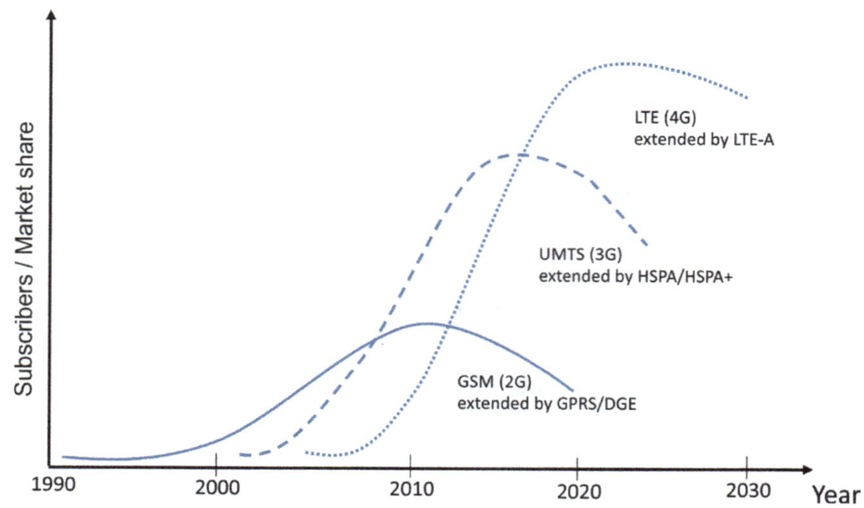

Figure 9. Relative adoption and life cycle of wireless technologies in the last three decades.

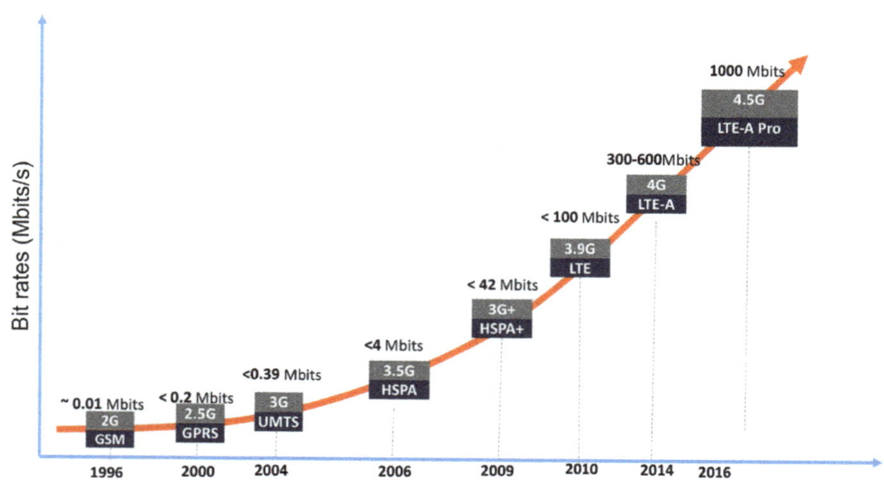

Figure 10. Throughput evolution of wireless technologies for access networks and WWAN.

- **The high number of users and no service guarantees in critical situations**: For some high priority smart cities applications, such as disaster and public safety management, the share of the same infrastructure with lower priority uses can compromise high priority applications requirements, thus making it ineffective.

- **Short wireless technologies life cycle**: Utilities are running their business models by considering equipments that last for more than 20 years. However, the modern mobile communications has a short life cycle, as depicted in **Figure 9**. In fact, Global System for Mobile Communications/General Packet Radio Service (GSM/GPRS, i.e. 2G/2.5G) networks is at the end of life, so that AT&T in the United States already started to switch off their GSM/GRPS infrastructure, as mobile network operators (MNOs) consider it wastes OPEX and spectrum efficiency [12]. This issue is also discussed for the old version of the Universal Mobile Telecommunications System (UMTS, i.e. 3G).
- **Short life cycle of electronics**: The traditional mechanical electricity meters, called also Ferrari meters, have a life cycle of about 25 years and more. The modern intelligent smart meters with all types of electronic components inside and its communication module will rarely reach this age of life cycle. As an example, the power utility in Italy ENEL has a total of more than 32 million of metering points and ERDF in France has more than 36 million meters. Considering this large number of end-devices, utilities must adjust their business models based on an intensive benefit–cost analysis.
- **Low battery**: Modern mobile networks are strongly consuming the power of the battery.
- **Indoor coverage and building penetration**: Most of the smart meters are placed in building basements, which are very rarely reached by the GSM coverage signal, especially in European cities. In this case, utilities must use a first level of communication, home area network (HAN) or building area network (BAN). This is also the problem for the communication in smart parking in the underground.

The above-cited challenges have pushed MNOs and equipment manufacturers to build associations, in order to unify their points of view together with industrials and to define the requirement of the next generation of wireless and mobile networks. The most known international association is the next-generation mobile network (NGMN) Alliance [13]. Their activities have resulted in new standards and new product and technologies of wireless networks, mostly known as low-power wireless area networks (LPWANs). These new generations and technologies build the focus of the next session.

5.4.2. *New wave of communication technologies*

As already stated in section 2.1, smart applications are already deployed using current technologies, and standards are forecasting more smartness in current deployment enabled by technologies and features to be proposed. As an example, LTE and LTE-A can give usable characteristics for smart city applications, in addition to LPWAN. IoT is tightly related to smart city applications as these applications rely for the most on a smart metering coupled with a pre- and/or post-intelligence. Thus, smart city LTE-enabled applications focus on IoT requirements. **Table 2** shows some LPWAN IoT characteristics, sufficient enough today for low rate applications.

Note that there exist wireless networks already used for low-power applications, such as bluetooth, Wi-Fi, and ZigBee. However, long-range performance and cellular M2M networks, on which smart cities are mostly based, would be costly and energy consuming, besides being expensive as far as hardware and services are concerned, while many connection devices are massively deployed in a smart city but need to send only small amounts of data over a long range when maintaining long battery life. In comparison with other technologies, LPWAN seat at the range shown in **Figure 11**. Following the use cases, other technologies can be used to the deployments of IoT.

Technology	SIGFOX	LoRa	NB LTE-M Rel. 13	LTE-M Rel. 12/13	EC-GSM Rel. 13	5G (under development)
Range (outdoor) (km) MCL (dB)	<13 160	<11 157	<15 164	<11 156	<15 164	<15 164
Spectrum Bandwidth	Unlicensed 900 MHz 100 Hz	Unlicensed 900 MHz <500 kHz	Licensed 7–900 MHz 200 kHz or Shared	Licensed 7–900 MHz 1.4 MHz or Shared	Licensed 8–900 MHz 2.4 MHz or Shared	Licensed 7–900 MHz Shared
Data rate	<100 bps	<10 kbps	<150 kbps	<1 Mbps	10 kbps	<1 Mbps
Battery life (years)	>10	>10	>10	>10	>10	>10
Availability	Today	Today	By end of 2016	By end of 2016	By end of 2016	By 2020

Table 2. Wireless technologies to build LPWAN for IoT connectivity.

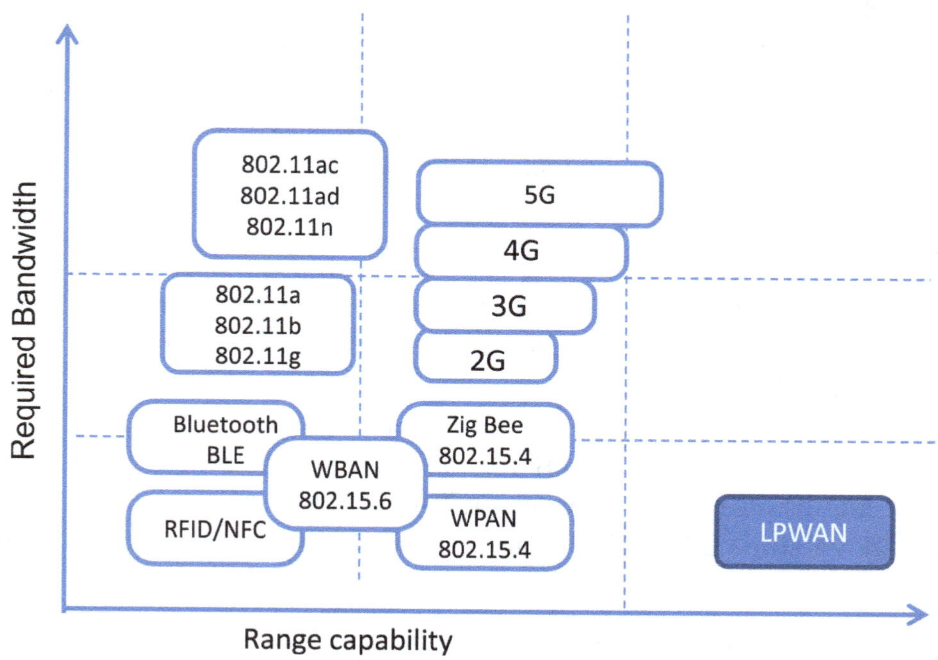

Figure 11. LPWAN range vs bandwidth in comparison with other technologies.

5.4.3. LPWAN

One of the major issues for M2M communications is the low-power long-range communication. LPWAN technologies, in order to achieve a long range capability, need to use high receiver sensitivities, up to -130 dBm compared with the -90 to -110 dBm in many traditional wireless technologies. This implies a higher energy per bit and thus a slower modulation rate. An example here is SigFox, which uses the extremely slow BPSK modulation. Some existing LPWAN platforms are described in the following:

- **LoRaWAN** is an LPWAN specification for wireless battery operated things with low data rate communicating over long distances, typically 15–20 km, promoted by LoRa Alliance12. It can be arranged to provide coverage similar to that of a cellular network, with millions of nodes and a battery life in excess of 10 years. In practice, LoRa networks are already deployed by cellular network operators who use existing masts to mount LoRa antennas, with possibility of combining antennas in some cases. Requirements for LoRAWAN meet target key requirements of IoT such as secure bidirectional communication, mobility, and localization services. The main advantage of this standard is the simplicity of local installation with an easiness of use for the end user and developer. Supported data rates range from 0.3 to 50 kbps, and the selection of the data rate is a trade-off between communication range and message duration.
- **SigFox** is a global IoT network designed exclusively for long range, small-message device communication with a low deployment cost. Historically, it is the first to have been developed. SigFox is a narrowband technology using BPSK modulation. It has thus the advantage of allowing the receiver to only listen in a tiny slice of spectrum, which mitigates the effect of noise. It has bidirectional functionality, but its capacity going from the base station back to the endpoint is constrained. Sigfox is currently deploying in many areas, even though it requires a new architecture because of the slow transmission.
- Other solutions existing in the market, such as *Weightless*, *Symphony Link*, *Nwave*, or *Ingenu*.

Note that most LPWAN solutions are addressing similar use cases identified within the 5G Umbrella, with the name of "Massive IoT". Complementarity with cellular can provide new opportunities to both technologies. However, some of previous solutions may be an alternative solution for smart cities while waiting for 5G standardization [14].

5.4.4. The 4.5 and 5G wireless shaped for smart cities applications

5.4.4.1. LTE and LTE-A

Following the requirements for smart cities applications, stated in section 4, these requirements meet for many applications the requirements of IMT-Advanced. LTE-A surpasses requirements of IMT advances. Actually, new users categories (cat1–6) are introduced starting from LTE release 8, with up to 20 MHz possible. LTE-A (Release 10) defines further categories, 6–8 and Release 12 defines LTE Cat-0 with a speed of 1 Mbps, and LTE-M defined in Release 13 with an even lower speed (kbps) are categories for machine type capabilities. LTE Cat-0 has been defined to suit needs of IoT applications and general M2M applications with much lower data rates, usually in short bursts, and treating only low levels. This new category has a reduced performance requirement, reduced complexity, and current consumption, while still with the LTE system requirements. In particular, peak downlink and uplink rate for Cat-0 is 1 Mbps, while Cat1–8 range from 10 to 1200 Mbps for downlink and from 5 to 600 Mbps for uplink. **Figure 12** illustrates the performance and complexity scaling in LTE releases in terms of data rates and bandwidth requirements.

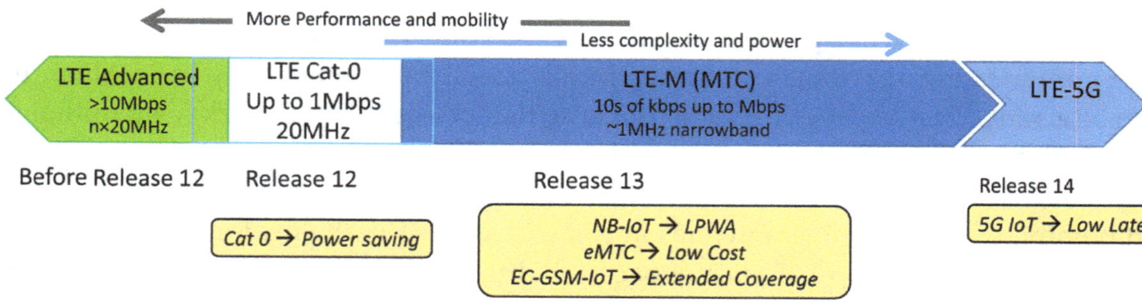

Figure 12. Scaling LTE-A following the requirements.

Note that Cat-1 chips are already suitable for many industrial applications including telematics, digital signage, and security systems. Nevertheless, Cat-0 chips, demonstrated, for example, by Sequans at CTIA's Super Mobility Week in Las Vegas, September 2015, are suitable for lower-speed applications. Similarly, Cat-M, already announced by Verizon, requires a new generation of highly optimized chipset design to reach the cost and market objectives [15]. The chipset makers are building dual mode chips supporting Cat-1 and Cat-M protocols at once, in order to allow the device to switch between modes following the conditions. In addition to UE categories, LTE/SAE (system architecture evolution) offers new solutions compared with earlier systems. In particular, the bandwidth in SAE can vary from 1.4 MHz up to 20 MHz, so the LTE/SAE can be used in various scenarios including IoT, LTE-A, and IEEE 802.16m (WiMAX 2).

5.4.4.2. 5G Enabling capabilities

Smart cities use distributed intelligence at different levels. Hopefully, IoT is a central part of 5G objectives. Compared with 4G, an increasing by a factor 1000 for mobile data volume is expected for 5G, the number of connected devices by a factor 100, guaranteed user data rate >50 Mbit/s, and a reduction by a factor 5 for the latency. The emerging technologies driving the 5G include, among others, device-to-device (D2D) technologies, software-defined networking (SDN), network functions virtualization (NFV), mobile edge computing (MEC), fixed mobile convergence (FMC). Below are some of the expected of enabling capabilities in 5G:

- **D2D**: Proximity-based device-to-device communications are an important 5G capacity enhancing technology. D2D UEs (DUEs) can act as transmission relays and set up multihop communication links, thus improving and extending network coverage. As for smart city applications, D2D communications will play an important role.

- **Fog Networking**: The large deployment in IoT applications implies a huge amount of data collected which, for some applications, needs to be centralized, and may use a cloud computing. This "cloud model" can be extended to the edge fog networking, which is a system-level horizontal architecture that distributes resources and services of computing, storage, control, and networking anywhere along the continuum from cloud to things. Ideally, IoT platform should fuse fog and cloud to have a device and service consolidation in addition to an optimization of lifecycle management of tenants and virtualized services, enhancement of data policy management, and integration of application and data management. A unified orchestration for fog and cloud can enforce service and security and integrate the entire IoT verticals. It is important to emphasize that the use of resource-rich fog servers running at the edge of mobile networks, known as mobile edge computing (MEC), as a part of the fog networking paradigm, and in conjunction with SDN and NFV, will be crucial in effecting low latency, high bandwidth and agility that will be able to connect trillions of devices.
- **C-RAN (Cloud RAN)**: This architecture promises the reduction in the cost of ultra-dense networks, expected especially with IoT deployments, through the simplification of the small base stations to remote antennas, possibly using a massive MIMO technology, and moving all the baseband processing to the cloud. This will improve the capacity by a joint processing of several signals from several remote antennas and can offer more optimization possibilities for coordinate multipoint (CoMP) and resource allocation. Networks can thus be flexibly and adaptively deployed to manage contents in a content centric way rather than the connection centric mindset currently adopted. The evolution toward C-RAN/ virtualized RAN (vRAN) architecture and interworking with mobile edge computing, the use of flexible, efficient, and reconfigurable hardware/software platforms, and data gathering and context information are key enablers for future IoT, leading to an evolution towards the "software defined everything."

6. Conclusions

The notion of smart city appeared in order to build future sustainable cities that benefit their inhabitants in a context of exponential urban growth, resource scarcity, demographic explosion, and strong environmental constraints. The smart cities is a set of applications, processes, and services that make use of engineering advancement to make an optimal use of the available infrastructures and foreseen the need in the future. The applications and processes of smart cities make use of massive data and information collecting, processing, and sharing. Therefore, a reliable communication and networking infrastructure should build the backbone of the smart cities, in order to make the data transmission possible. A large number of applications build the application layer of the smart city architecture, where each application has its own ICT requirement and expectation. Therefore, building such ICT infrastructure will involve different technologies depending on the application and the deployment environment. Wireless technologies are the most desired solution, because of all the economic and societal benefits they promise, particularly flexibility and ease of deployment. However, there are still challenges that wireless communication has to cope with, like power consumption, ease of installation, great indoor coverage. To overcome these challenges, equipment manufacturers and mobile network operators have unified their activities to develop and deploy a new wave of wireless technologies, named low-power wireless area network.

Author details

Abdelfatteh Haidine*, Sanae El Hassani, Abdelhak Aqqal and Asmaa El Hannani

*Address all correspondence to: a.h.haidine@ieee.org

Information Technologies Laboratory (LTI), National School of Applied Sciences, El Jadida, Morocco

References

[1] The Department of Economic and Social Affairs of the United Nations (DESA). World Economic and Social Survey 2013-Sustainable Development Challenges. Annual report, E/2013/50/Rev. 1 (ST/ESA/344). 2013

[2] United States Habitat. Hot Cities: Battle-ground for Climate Change—Cities and Climate Change. Global Report on Human Settlements. 2011

[3] ITU-T Focus Group on Smart Sustainable Cities (FG-SSC). An Overview of Smart Sustainable Cities and the Role of Information and Communication Technologies. Focus Group Technical Report. 2014

[4] Dameri R, Rosenthal-Sabroux C. Smart City—How to Create Public and Economic Value with High Technology in Urban Space. Springer International Publishing, New York, NY; 2014.

[5] Celino I, Kotoulas S. Introduction to Smart Cities. IEEE Internet Computing. 2013;17(6): 8–11.

[6] Geisler K. (IEEE Smart Grid Newsletters). The Relationship between Smart Grids and Smart Cities [Internet]. May 2013. Available from: http://smartgrid.ieee.org/newsletters/may-2013/therelationship-between-smart-grids-and-smart-cities [Accessed: 10.06.2016].

[7] Höller J, Tsiatsis V, Mulligan C, Karnouskos S, Avesand S, and Boyle D. From Machine-to-Machine to the Internet of Things—Introduction to a New Age of Intelligence. Elsevier, Amsterdam; 2014.

[8] Meters and More Alliance. Meters and More—Open technologies [Internet]. Available from: http://www.metersandmore.com/technology/ [Accessed: 10.06.2016].

[9] PRIME Alliance. Available from: www.prime-alliance.org/ [Accessed: 10.06.2016].

[10] G3-PLC Alliance. Available from: www.g3-plc.com/ [Accessed: 10.06.2016].

[11] Sauter M. From GSM to LTE-Advanced: An Introduction to Mobile Networks and Mobile Broadband. 2nd ed. Wiley, Hoboken, NJ; 2013.

[12] AT&T. Frequently Asked Questions Regarding 2G Sunset [Internet]. 2016. Available from: https://www.business.att.com/content/other/2G-Sunset-FAQ_2016.pdf [Accessed: 10.06.2016].

[13] Next Generation Mobile Network Alliance (NGMN). About the NGMN Alliance—Status and 5G Work Programme [Internet]. April 2016. Available from: http://www.ngmn.org/fileadmin/ngmn/content/documents/pdf/about_us/1604_NGMN_Alliance_Overview.pdf [Accessed: 10.06.2016].

[14] Orange. Orange deploys a network for the Internet of Things [Internet]. September 18, 2015. Available from: http://www.orange.com/en/Press-and-medias/press-releases-2016/press-releases-2015/Orange-deploys-a-network-for-the-Internet-of-Things [Accessed: 10.06.2016].

[15] RCR Wireless News—Intelligence on all things wireless. Verizon chip partner first to announce lower cost Cat-M solution for IoT [Internet]. February 16, 2016. Available from: http://www.rcrwireless.com/20160216/internet-of-things/lte-iot-sequans-cat-m-verizon-tag4 [Accessed: 10.06.2016].

10

Control Strategies for Smart Charging and Discharging of Plug-In Electric Vehicles

John Jefferson Antunes Saldanha,
Eduardo Machado dos Santos,
Ana Paula Carboni de Mello and
Daniel Pinheiro Bernardon

Abstract

This chapter aims to provide an overview of the plug-in electric vehicle (PEV) charging and discharging strategies in the electric power system and the smart cities, as well as an application benefiting both consumers and power utility. The electric vehicle technology will be introduced. Then, the main impacts, benefits and challenges related to this technology will be discussed. Following, the role of the vehicles in smart cities will be presented. Next, the major methods and strategies for charging and discharging of plug-in electric vehicles available in the literature will be described. Finally, a new strategy for the intelligent charging and discharging of electric vehicles will be presented, which aims to benefit the consumer and the power utility.

Keywords: plug-in electric vehicles, smart charging and discharging, control strategies, smart grid, vehicle-to-grid

1. Introduction

The automotive industry is growing rapidly, and the vehicles powered by fossil fuels make up the largest share of this sector. However, environment concerns have also increased considerably, raising questions about the harmful effects of pollutants emitted by these vehicles.

The plug-in electric vehicles (PEVs) represent an alternative to reduce the emission of pollutants from vehicles powered by fossil fuels. These vehicles are propelled by electric motors powered by the energy stored in their batteries. Thus, to charge the battery, it is necessary to connect it to an electrical outlet.

Despite the charging need, a PEV can also be used to provide ancillary services to the power system, aiming the grid benefit. Such vehicles are abbreviated as V2G, from vehicle-to-grid, and, as major examples, are able to provide regulation, renewable sources support and distribution losses minimization.

Putting a car to charge its battery will cause it to act as a load to the power grid. Then, if a significant amount of electric vehicles starts to request charging at the same time, the system may suffer certain complications. Given so, it is important to make a coordinated charging control of the electric vehicles.

With the previous approaches that are further discussed, the PEV plays an important role in smart cities. In this sense, this chapter presents an overview of the PEV charging and discharging strategies in the electric power system and the smart cities. An application that benefits both consumer and power utility is presented as well. In addition, the main impacts and benefits brought up from the PEV charging and discharging are discussed. The challenges associated to this technology are showed.

First, the electric vehicle technology is presented, and distinction and comparison between major types are made. In the following section, the main impacts, benefits and challenges related to this technology are presented. Furthermore, the role of plug-in electric vehicles in smart cities is exposed. Then, the major methods and strategies for the PEVs charging and discharging are presented. Finally, a new strategy for smart PEV charging is presented, which aims to benefit both the consumer and the power utility.

2. Electric vehicle technology

The EVs are vehicles propelled by electric motors, the latter being powered by electrical energy stored in energy storage devices, such as batteries and supercapacitors. Generally, they are classified into three distinct groups: battery electric vehicle (BEV), hybrid electric vehicle (HEV) and plug-in hybrid electric vehicle (PHEV). A broader classification, called plug-in electric vehicles (PEVs), is commonly used to comprise the BEVs and PHEVs, because both types need to connect to the power grid to charge their batteries.

Battery electric vehicles, also known as purely electric, are powered by electric motors that receive electricity stored in a battery. The battery is charged through a power grid connection or through replacing it for another one already charged. The HEVs combine the previous vehicles with conventional ones powered by internal combustion engines (ICEs), with the first allowing higher efficiency and fuel economy, and the latter providing greater driving autonomy. The batteries of these vehicles are charged by regenerative brake and by the combustion engine. Finally, the plug-in hybrid electric vehicle has the same electric and combustion motors combination of the HEV, but the battery can be charged by connecting the PHEV in the power grid in addition to regenerative braking and combustion engine [1].

In terms of energy consumption [2], it was proved that BEVs and HEVs provide higher fuel economy when compared to internal combustion vehicles, including the case when the electricity is generated from petroleum-derived resources. Besides that, the use of HEVs and BEVs can lead to greenhouse gas emission reduction. Despite the fuel economy increase, the global reliability of HEVs and PHEVs is lower when compared to combustion engine vehicles, as the architecture of the first two is more complex than the latter, given that numerous subsystems are involved [3]. **Table 1** shows some comparisons between BEV, HEV and PHEV, including: the propulsion engine type that they have, the kind of energy storage system and from where it comes, and the major aspects that differentiate them and the main problems faced.

Characteristics	BEV	HEV and PHEV
Propulsion	• Electric motor	• Electric motor • Internal combustion engine (ICE)
Energy storage subsystem	• Battery • Supercapacitor	• Battery • Supercapacitor • Fossil or alternative fuels
Energy source and infrastructure	• Power grid charging facilities	• Gas station • Power grid charging facilities (for PHEV)
Main observations	• Zero local emissions • High energy efficiency • Fossil fuel independent • Short driving range • High initial price	• Zero local emissions • High fuel economy • Fossil fuel dependent • Long driving range • Higher cost than ICE vehicles
Main issues	• Battery storage capacity • Charging stations • Cost • Battery lifetime	• Battery storage • Control, optimization and management of different energy sources

Adapted with permission from Chan et al. [4].

Table 1. Major comparisons between BEV, HEV and PHEV.

By not emitting locally harmful gases, the purely electric vehicles appear as an ideal solution for dealing with the energy crisis and global warming. However, due to the high initial investment, low driving range and long charging time, the BEVs still show themselves limited for large commercial incorporation. The HEVs are presented as a viable option to overcome the previous limitations, having expressive autonomy, but excessively emitting pollutants. Thus, seeking to maximize the driving autonomy of the vehicle along with minimizing pollutant emissions, PHEVs appear as the best alternative, overlapping the HEVs.

Combining the benefits and prospects mentioned above, the PEVs are presented as a new highly promising technology with great anticipation to play a major role in the power and automotive industries in the forthcoming decades. However, attention should be paid to the fact that the PEV can act both as a load and power source to the power grid. The first case occurs when the PEV is charging its batteries, that is, consuming power from the grid. The second case can happen with the vehicle communicating with the grid to act on its benefits, either providing active energy or controlling its charging rate. This type of concept is called vehicle-to-grid or V2G.

3. The PEV role in smart cities

The urban energy systems modernization, combined with the current requirements to meet the demand growth, contributes to the creation of a new approach for managing the electric power system, called smart grids. The smart grids seek to address modern energetic challenges in a city, using information technology (IT), automation and communication. In this way, the optimization of the power grid operation is aimed, allowing integration of equipments to the grid and including alternative sources of electric power generation, in addition to online data gathering of energy demand and supply to manage peak loads.

The integration of PEVs in the electricity distribution network is one of the major technological developments brought by smart grids, as it provides the application of charging and discharging control strategies in the vehicles to achieve a given goal.

Among the main benefits of the charging and discharging of PEVs, the following are highlighted:

- Greater integration of the PEVs in the power grid
- Reduction of pollutants in cities
- Cost reduction
- Smart charging according to consumer's need
- Power losses minimization
- Peak load reduction
- Profit maximization
- Energy demand management
- Energy resources management

- Greater integration of renewable energy sources
- Higher reliability and efficiency of the power system
- Power system operation optimization
- Among others

4. Impacts and benefits of PEV charging and discharging and challenges that prevent this technology diffusion

This section presents the key impacts, benefits and challenges that prevent the diffusion of the plug-in electric vehicle technology. In the first section, the impacts that the PEV charging without coordination can cause to the power grid are presented. Following, the main benefits originated from the vehicles, in environmental terms, ancillary services that can be provided to the grid and other benefits are exposed. Finally, the third section presents the key challenges linked to the low implementation of the vehicles under study.

4.1. PEV charging impacts

The plug-in electric vehicle charging is cyclical, variable and somewhat unpredictable, as it acts as a load to the power grid. If a significant amount of PEVs requests energy at the same time and if a proper coordination is not used, the charging process may randomly occur both in time and in space. This may bring certain complications to the electrical grid, particularly at peak periods. This is a very likely assumption, since the PEV owners could arrive at home after work and begin to charge the vehicles during a period of high demand.

A typical PEV under charging more than doubles the average load of a household, drastically changing the consumer's load curve. In addition, the PEV charge without coordination can increase the load at peak times and cause local problems in distribution networks, as well as increase energy losses and voltage deviations, compromising the power quality [5–9]. Another aspect is that they can also lead to conductor and distribution transformer overload and reduce the grid reliability and stability.

In the financial context, an increase in consumers' electricity bill is pointed out, due to the demand growth caused by the vehicle charging process, as well as capital expenditures, resulted from the need to reinforce the grid and also from the energy loss costs [10, 11].

The plug-in electric vehicle also affects the electrical power system reliability and efficiency. For instance, when requesting charging during peak hours, some fast response power plants with relatively low efficiency can be activated to meet the growing demand [12]. Therefore, the generation system global efficiency decreases. In addition, the reliability of these power plants is also low, reducing the overall system reliability.

Briefly, PEV charging without proper coordination can increase the consumers' average load, causing undesirable consumption peaks. Furthermore, it can overload transformers and violate transmission lines thermal limits. In addition, it can cause the dispatch of expensive and highly polluting power plants to meet the growing demand. Voltage deviations and reduction in reliability and efficiency may also arise. Finally, the electricity bill may increase and the system stability may be committed.

However, despite the aforementioned impacts, PEVs can also aid the electric power system through the implementation of effective measures to manage the loading. The following section presents these benefits from an environmental point of view and the possible advantages for the power grid.

4.2. PEV benefits

Plug-in electric vehicles can also create opportunities to the electricity sector if a planned and coordinated control strategy is applied. The benefits can be in environmental terms, through ancillary services provision and other benefits not included in the foregoing. The following three subsections feature, respectively, these benefits according to the literature.

4.2.1. Environmental benefits

CO_2 emissions can drop dramatically if the PEVs replace the conventional internal combustion vehicles. In addition, the use of PEVs can directly reduce the emission of greenhouse gases. The analysis made by Ramteen and Denholm [13] shows that the PHEV can lower pollutant emissions associated with the automotive industry and, adopting the V2G concept, these reductions can be much more significant. Much of this decrease is dictated by the conditions in which the vehicle will be charged, since emissions can be significantly reduced if pollutant sources are used in power generation.

In scenarios with low, medium and high penetration of PEVs, a significant reduction in greenhouse gases can be achieved, lowering the automotive industry reliance on fossil fuels. The work by Samaras and Meisterling [14] obtained a 32% greenhouse gas reduction with the inclusion of PHEVs, compared to conventional ones. However, if the electric power generation is not derived from a source with low pollution rate, this reduction becomes short.

It is evident that the plug-in electric vehicle adoption will significantly reduce the pollutant emission into the atmosphere when compared to traditional vehicles powered by fossil fuels. However, one should be careful when comparing the global emissions reduction that includes pollution generated from power plants used to meet the demand. In this case, the reduction may not be as significant, and even worse in some scenarios. Allied to these reductions, PEVs can also benefit the electric system through the provision of ancillary services, as will be seen in the next subsection.

4.2.2. Ancillary services

In electric power systems, ancillary services are the services necessary to maintain the system reliability and the balance between supply and demand and also aid the operation. Among these services, the frequency control, voltage control, spinning reserve, peak load leveling, assistance in starting emergency generators (black start), support for reactive power and active loss compensation are highlighted.

The plug-in electric vehicle combined in the V2G concept can work for the grid benefit by providing various ancillary services. However, a single electric vehicle unit has limited capacity to represent a significant benefit to the grid. Thus, the PEVs are usually grouped in fleets or, as [15] defines and explains, in aggregators.

The aggregator is an entity that herds several PEVs to generate a significant, beneficial and effective impact in the power grid, acting as an interface between the system operator, usually the distribution management system (DMS), and the fleet connected. Therefore, it enables the vehicle participation in the electricity market. It behaves as a decision-maker, having an optimized strategy to effectively act as generation/storage, being capable to provide ancillary services to the grid or to operate as a controllable load to be charged in the most beneficial manner for the system.

The basic scheme of communication between the aggregator and the electric power system is illustrated in **Figure 1**, where the electricity flows in a single direction from generating plants to consumers. Between the PEVs and the aggregator, the energy flow is bidirectional, the latter having control over the charging or discharging of the first. The ancillary services necessary for the grid and that can be performed by the PEV fleet are remotely requested by the system operator, characterizing the energy and communication flows in both directions.

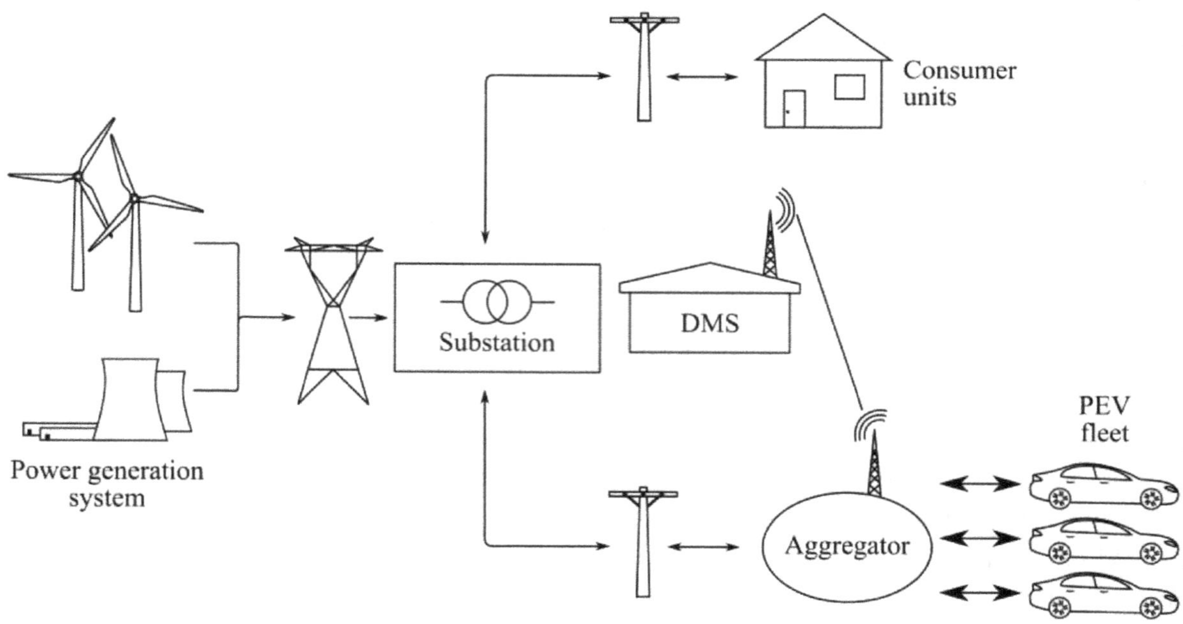

Figure 1. Illustrative scheme of the communication between the aggregator and the electrical power system [16].

By the V2G concept, the PEVs can be used to provide the following ancillary services for the electric power system:

- Frequency and voltage regulation [16, 17]
- Transient stability improvement [18]
- Peak shaving [19, 20]
- Valley filling [19]
- Spinning reserve [20, 21]
- Power flow optimization [20]
- Power quality improvement [20]

Briefly, plug-in electric vehicles can provide ancillary services of frequency and voltage regulation through V2G concept, and they have been proved to be economically feasible. Furthermore, they can also be used to improve transient stability, manage peak load, fill the consumption load valleys, serve as spinning reserves, optimally manage the power flow and improve the power quality. However, in order to have a share in the market and provide significant services, vehicles are usually concentrated in aggregators, which act as an interface between the fleet and the system operator.

In addition to the already mentioned benefits, the next subsection introduces a modern and undergoing study topic about the use of electric vehicles to support the renewable energy entry and other benefits that are not considered as ancillary services.

4.2.3. Further benefits

The use of EV fleet may considerably increase the demand forecast accuracy. To do so, the vehicle fleet acts as a variable power storage system, absorbing prediction errors and, consequently, reducing the cost to be paid for them. This type of benefit is only achieved from a certain EV penetration level [22].

By combining the photovoltaic solar energy generation with the PEV charging, it is possible to maximize the benefits and minimize the costs, through high penetrations of the two technologies in the electricity sector, what reduces emissions from the EV charging at peak times. The first provides power during peak midday in summer, reducing the need for additional generation capacity. The second absorbs energy from photovoltaic plants that would be wasted due to low demand in the spring [23, 24].

The V2G concept increases the flexibility of the grid to a better use of intermittent renewable energy resources, acting in a way to increase the demand forecast accuracy as a variable source of storage.

The PEV benefits to the electric power system have been proven by several studies. However, some challenges are still imposed to the full accomplishment of the benefits. Next, some of the challenges that prevent the PEVs and V2G concept diffusion are presented.

4.3. Challenges of the V2G concept

Despite the many benefits from the PEV application in the V2G concept, some impediments and barriers still hinder its implementation, as shown below.

In terms of the vehicle battery, its use to provide ancillary services will reduce its lifespan due to the extra charge and discharge cycles. To estimate a cost related to this degradation is an interesting feature to incorporate in the economic viability calculus, but this is a complex task, given that the technologies are still under development. Reference [15] states that through intelligent control, charging time and optimal energy flow it is possible to minimize the additional battery degradation rate because of ancillary services. Still, there is a major barrier to achieve this intelligent control.

A secure communication among the aggregator, a large number of PEVs and the system operator must be ensured, where a reliable bidirectional infrastructure must exist. In addition, sensors and smart metering should be installed. The charging energy growing demand may require, in some cases, expansion of the generation capacity, as mentioned previously, since the PEV charging process may cause considerable impact on distribution equipments, overloading them. All these points converge to the need for significant investments in the electrical power system to meet the changes.

Other commonly presented challenges that prevent the PEV diffusion are as follows: the high initial investment cost compared with conventional vehicles, their low autonomy, the resistance that the automotive and oil industries may offer, as well as consumer acceptance of this new technology.

5. Strategies and methods for PEV charging and discharging

The plug-in electric vehicle charging can be coordinated or uncoordinated. In uncoordinated charging, there is no control over the charge rate of the PEVs. They begin to charge immediately when plugged in or after a predetermined time and only cease charging when the battery is fully charged or the vehicle is disconnected from the grid. The coordinated charging applies a type of strategy that controls the PEV charging.

Coordinated charging control can only be done to limit or define the PEV charging rate or through active power injection. The first case is characterized by unidirectional power flow (from the grid to the vehicle), and the second case is characterized by bidirectional flow (from the network to the vehicle and vehicle to the network). It is through coordinated charging that the impacts and benefits of the previous section can be mitigated and availed.

The unidirectional power flow is only used to charge the PEV battery, having a low cost of implementation. In this regard, the battery presents no additional degradation, given that it does not discharge by supplying power to the grid. For this type of flow, only one electrical connection to the power grid is required. Among the main benefits, it includes the possibility of applying a simple control for coordination, which facilitates the mitigation of impacts on the network, achieved by limiting the charging rate.

The bidirectional power flow involves higher costs than the previous one, but can be used both for charging control and supplying energy to the grid when required. However, due to the discharge cycles, the battery undergoes additional degradation. For its implementation, it is necessary to have bidirectional communication and smart metering deployed. Thus, this type of power flow allows providing ancillary services to the grid and renewable energy sources entry support. **Table 2** lists some points of comparison between unidirectional and bidirectional power flow.

	Unidirectional power flow	Bidirectional power flow
Power flow	• Power flow in one direction • Only PEV charging	• Power flow in two directions • PEV charging and discharging
Cost	• Low	• High
Battery effect	• No extra degradation	• Extra degradation due to battery discharge
Electrical distribution system	• No need for investment and modernization	• Investment and modernization required
Requirements and challenges	• Power grid connection	• Bidirectional communication and power grid connection • Smart and suitable metering and sensors • Considerable information exchange • Extra cost and investments • Energy losses • Device stress
Benefits	• Simplify power grid connections • Simple control and management • Provide services based in the adjustmentof charging rates • Supplies or absorbs reactive energy without discharging the battery	• Ancillary services • Active power regulation and voltage and frequency stabilization • Spinning reserves • Reactive power support • Peak shaving • Valley filling • Energy balance • Harmonic filtering • Renewable sources entry support

Adapted with permission from Yilmaz and Krein [25].

Table 2. Comparisons between unidirectional and bidirectional power flow.

The coordinated charging strategies can be decentralized or centralized. In the decentralized or distributed charging, the PEV owner decides when he will charge the battery. In the centralized, a central entity gathers information about the grid, load and generation, being able to decide and control the charging or discharging of a PEV fleet. **Figure 2** shows a flow chart with the strategies that descend from the PEV charging.

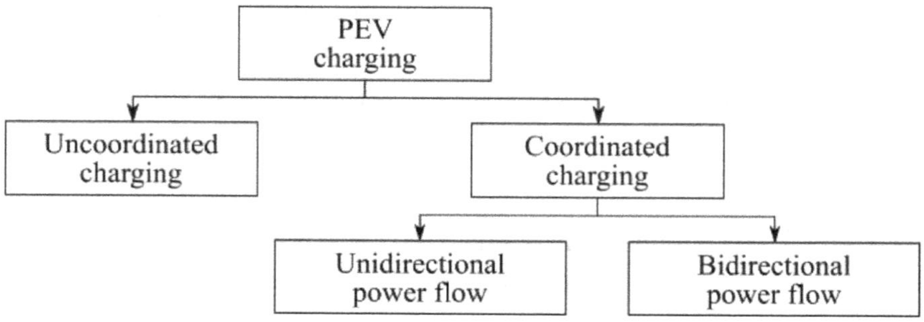

Figure 2. Different strategies arising from PEV charging and discharging.

The two following sections present some methods for uncoordinated and coordinated PEV charging. Each method is based on certain assumptions regarding the available infrastructure, charging rates and duration, status, size, technology and energy capacity of the battery, PEV type and mathematical approaches.

5.1. Uncoordinated charging

Strategies to uncoordinated charging are scarce, given the facts that do not result in significant benefits for the grid. They are directly dictated by the users, not being possible to carry out coordination and control to act in favor of the electric power system.

The first uncoordinated charging strategy happens when the PEVs start to charge immediately when connected or after a certain delay set by the owner. In the first case, the user has no incentive or information necessary to make a smart charging in favor of the power system. In the second case, he has information and the delay allows the charging to happen in periods outside the peak time. This strategy often happens because users come home after work and connect their vehicles to charge so the battery is fully charged the next day [26].

Even in an uncoordinated way, but with an incentive for the PEV owners to take action, it is common to use two electricity prices. A more expensive one is applied during peak periods and a cheaper off-peak. Thus, it is assumed that the users of the double tariff will charge the vehicles during low demand period where the value to be paid for the energy is lower.

It is noteworthy that with this type of charging strategy there is no control of the charging rate and much less of the energy flow from the vehicle to the grid. Impact minimization can be achieved by encouraging the PEVs to charge during off-peak periods through the use of differentiated tariffs. Coordinated charging is the solution to this problem, and the next section presents some of the major strategies, methods and approaches available in the literature.

5.2. Coordinated charging and discharging

The researches in methods and strategies of coordinated charging and discharging for the power grid benefit have grown significantly in recent years, each one considering different approaches with diversified assumptions. Some of the most relevant alternatives are presented below, separated into two subsections. At first, unidirectional power flow is considered and secondly, the two-way flow is taken into account.

5.2.1. Unidirectional power flow

Among the main objectives adopted for PEV coordinated charge with unidirectional power flow, the following are highlighted:

- Minimize energy losses [26–28]
- Frequency regulation [29]
- Maximize aggregator's profit [30, 31]
- Minimize charging cost [28, 32–35]
- Reduce voltage deviations [36]
- Reduce power grid overload [36]
- Minimize the deviation between energy bought in the market and energy consumed by the PEVs [37]
- Prevent the electricity distribution network congestion [38]
- Reduce peak load [31, 39]
- Maximize energy delivered to the PEVs [28]

The major techniques used to solve the coordinated charging problem with unidirectional power flow are as follows:

- Quadratic programming [26]
- Dynamic programming [26, 29, 40]
- Heuristic algorithm [30]
- Optimized selection [30]
- Linear programming [32, 38]
- Search techniques [33]
- Neural networks [33]
- Artificial immune systems (AISs) [36]
- Nonlinear programming [27]

- Evolutionary programming [34]
- Fuzzy linear programming [31]
- Fuzzy discrete particle swarm optimization [28]
- Fuzzy genetic algorithm [28]

5.2.2. Bidirectional power flow

The main objectives addressed by authors for the PEV charging and discharging considering the bidirectional power flow are as follows:

- Minimize energy costs [41–43]
- Power regulation [41]
- Improve the photovoltaic generation predictability [44]
- Maximize the satisfaction of vehicle owners [45]
- Minimize the electric power system operational cost [45, 46]
- Limit carbon emissions [46]
- Minimize power mismatch on the grid [47]
- Reduce peak demand [48]
- Increase the electrical system reliability [43]
- Frequency regulation [49]

Among the techniques used to solve the charging and discharging problem with bidirectional power flow, the following are listed:

- Sequential quadratic programming [41]
- Particle swarm optimization [44]
- Mixed-integer quadratic programming [42]
- Mixed-integer linear programming [46]
- Linear programming [47]
- Self-adaptive evolutionary algorithm based on symbiotic organism search [43]

In the simulations, the profiles and load characteristics due to PEV insertion were defined in various ways. Among them, it is highlighted the use of fuzzy expert system, deterministic programming, stochastic programming, mixed-integer quadratic programming, Monte Carlo simulation and fuzzy C-means clustering.

The analyzed methods approached both centralized and decentralized charge and discharge. Different interfaces were created to relate with the PEVs, such as the aggregator, the distribution system operator and each vehicle unit.

6. A new charging control strategy considering consumer's will

The PEV will only be used to assist the power grid if the consumer is willing to provide the automobile for this purpose. Therefore, besides taking into account the current condition of the power grid, one must consider the consumer's requirements to have the vehicle fully charged within the connection time.

Thus, it is defined that the vehicle owner must enter a priority in which he wants the battery to be fully filled within the connection time to charge. In this context, a low value indicates that he is willing to provide its PEV to the grid and the battery may not be fully charged after disconnection. A high priority ensures full charge, not providing the vehicle to benefit the power system. As this is information set by the consumer and it is uncertain to the system, the priority insertion is performed by applying a fuzzy logic-based system, given that it allows the incorporation of human knowledge in mathematical models.

This priority entered by the consumer is the major contribution of this strategy, as the studies available in the literature do not consider this type of information in the charging rate control.

Considering the two centralized and decentralized approaches of the coordinated charging as defined before, this study represents a combination of both. The centralized approach is considered because the aggregator gathers information about the grid, load and generation, and sends command to each PEV. The second approach is taken into account by the fact that it includes the will of the PEV owner to have a fully charged battery according to his need.

The unidirectional power flow was verified to be the most feasible when considering the battery degradation, so this type of flow is considered. In addition, the implementation of two-way power flow is further than the one-way, given the fact that the grid will have to be improved and the electricity market will have to undergo adjustments.

The optimization goal is to minimize the energy losses in the power grid. The applied optimization technique to solve this problem is the heuristic search for generating and testing or the exhaustive search.

The proposed system has two main interfaces: the distribution management system (DMS) and the aggregator. The aggregator is responsible for gathering PEVs and consumers' information, receiving a power value from the DMS to distribute between vehicles meeting the priorities. The DMS acts to meet the requirements of all aggregators under its command, determining a power value that optimizes the operation of the network.

The general understanding of the proposed charging control system for the benefit of the electric power system can be extracted from **Figure 3**, illustrating the exchange of information between the DMS and an aggregator.

Once connected to the network for charging, the PEV reports to the aggregator with the following data: (1) its initial charge (SOC_i), representing the charging state of the battery at the moment; (2) the connection starting time (t_i), indicating the time that it was connected to charge and (3) the battery capacity B_{cap}, that is the energy the battery can store.

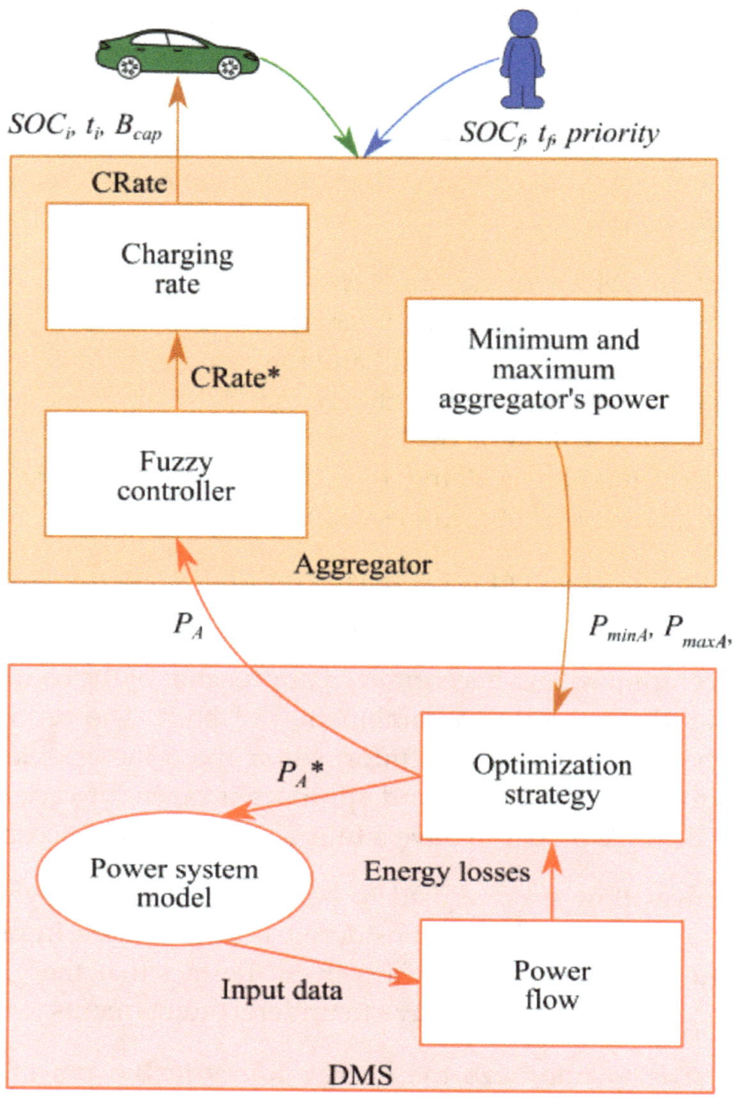

Figure 3. General architecture of the proposed charging control system considering the consumer's will. *CRate* – charging rate; SOC_i – initial state of charge; SOC_f – final state of charge; t_i – PEV connection time; t_f – PEV inserted disconnection time; B_{cap} – battery capacity; P_{minA} – minimum aggregator power; P_{maxA} – maximum aggregator power; P_A – available power to aggregator.

Still in the connection, the vehicle owner also informs three data to the aggregator: (1) the final state of charge (SOC_f) that he wants the battery to be charged; (2) the time he wants the battery to be charged (t_f) and (3) the *priority* in which he wants its battery to be fully charged to the final state of charge in the connection time to the grid.

Having received the six input variables from each vehicle in the fleet, the aggregator determines the minimum and maximum powers that should be received from the system to meet a minimum or total percentage of the vehicle charges in the stipulated time.

Each aggregator notifies the DMS with their minimum and maximum powers, respectively, P_{minA} and P_{maxA}. The DMS that has information about the electric power system, along with the aggregators' power, performs the energy distribution searching for an optimal way to minimize or maximize a certain goal.

The DMS returns the optimal power calculated to each aggregator within the range limited by the latter. The aggregator, based on a fuzzy controller acting on its interface, calculates each PEV charging rate, returning the value so that the vehicles can be charged.

The index * in the variables P_A and *CRate* indicates the initial and intermediate values for both the power available to the aggregator and charging rate. They are used in the proposed system but are not the final values to be returned. That is, P_A^* and *CRate** are part of the calculation, but only P_A is passed on to the aggregator to be distributed among the PEVs and only *CRate* indicates with which rate each vehicle will charge.

7. Conclusions

This chapter presented the importance of plug-in electric vehicles for the electric power system and the smart cities. Faced with the characteristics presented, the main strategies and methods of PEV charging and discharging control available in the literature were showed.

Out of the different methods searched, it was verified that they are based on certain assumptions regarding the available infrastructure, PEV type, rates, charging duration, status, size, technology and energy capacity of the battery, as well as mathematical approaches. Among the methods, more attention is being given to the coordinated charging with unidirectional power flow, and the same is closer to be currently widely applied in smart grids and smart cities. This occurs given the fact that the method only controls PEV charging rate seeking benefits, such as reducing pollutants or cost. In addition, the current infrastructure does not need significant reinforcements.

Finally, a new strategy for PEV smart charging is presented, where the advantages of the application and its role in smart cities are presented. The proposed methodology aids aggregators to reduce the impact on the grid, since the charging is carried out in a planned and controlled way. Application of this methodology can also help evaluating how electric vehicles can contribute to ancillary services, and which of these services would be most appropriate for the participation of PEVs.

Acknowledgements

The authors would like to thank the technical and financial support of Coordination for the Improvement of High Level Personnel (CAPES) and the National Center of Scientific and Technological Development (CNPq).

Author details

John Jefferson Antunes Saldanha[1*], Eduardo Machado dos Santos[1], Ana Paula Carboni de Mello[1] and Daniel Pinheiro Bernardon[2]

*Address all correspondence to: john.jefferson.as@gmail.com

1 Federal University of Pampa, Alegrete, Rio Grande do Sul, Brazil

2 Federal University of Santa Maria, Santa Maria, Rio Grande do Sul, Brazil

References

[1] Chan C. An overview of electric vehicle technology. Proceedings of the IEEE. 1993;81(9): 1202-1213. DOI: 10.1109/5.237530

[2] Williamson S, Emadi A. Comparative assessment of hybrid electric and fuel cell vehicles based on comprehensive well-to-wheels efficiency analysis. IEEE Transactions on Vehicular Technology. 2005;54(3):856-862. DOI: 10.1109/TVT.2005.847444

[3] Masrur M. Penalty for fuel economy-system level perspectives on the reliability of hybrid electric vehicles during normal and graceful degradation operation. IEEE Systems Journal. 2008;2(4):476-483. DOI: 10.1109/JSYST.2008.2005714

[4] Chan C, Bouscayrol A, Chen K. Electric, hybrid, and fuel-cell vehicles: architectures and modeling. IEEE Transactions on Vehicular Technology. 2010;59(2):589-598. DOI: 10.1109/TVT.2009.2033605

[5] Ipakchi A, Albuyeh F. Grid of the future. IEEE Power and Energy Magazine. 2009;7(2): 52-62. DOI: 10.1109/MPE.2008.931384

[6] Masoum A, Moses P, Hajforoosh S. Distribution transformer stress in smart grid with coordinated charging of plug-in electric vehicles. In: IEEE, editor. IEEE PES Innovative Smart Grid Technologies; 16–20 Jan. 2012; Washington, DC. Washington, DC: IEEE; 2012. p. 1-8. DOI: 10.1109/ISGT.2012.6175685

[7] ElNozahy M, Salama M. A comprehensive study of the impacts of PHEVs on residential distribution networks. IEEE Transactions on Sustainable Energy. 2014;5(1):332-342. DOI: 10.1109/TSTE.2013.2284573

[8] Shafiee S, Fotuhi-Firuzabad M, Rastegar M. Investigating the impacts of plug-in hybrid electric vehicles on power distribution systems. IEEE Transactions on Smart Grid. 2013;4(3):1351-1360. DOI: 10.1109/TSG.2013.2251483

[9] Zhang C, Chen C, Sun J, Zheng P, Lin X, Bo, Z. Impacts of electric vehicles on the transient voltage stability of distribution network and the study of improvement

measures. In: IEEE, editor. IEEE PES Asia-Pacific Power and Energy Engineering Conference; 7–10 Dec. 2014; Hong Kong. Hong Kong: IEEE; 2014. p. 1-6. DOI: 10.1109/APPEEC.2014.7066085

[10] Halbleib A, Turner M, Naber J. Control of battery electric vehicle charging for commercial time of day demand rate payers. In: IEEE, editor. 2012 IEEE PES Innovative Smart Grid Technologies, ISGT 2012; 16–20 Jan. 2012; Washington, DC. Washington, DC: IEEE; 2012. p. 1-5. DOI: 10.1109/ISGT.2012.6175728

[11] Veldman E, Verzijlbergh R. Distribution grid impacts of smart electric vehicle charging from different perspectives. IEEE Transactions on Smart Grid. 2015;6(1):333-342. DOI: 10.1109/TSG.2014.2355494

[12] Giglioli R, Giuntoli M, Lutzemberger G, Poli D. Impact of a large fleet of EVs on the efficiency and reliability of an electric power system. In: IEEE, editor. 2014 IEEE International Electric Vehicle Conference (IEVC); 17–19 Dec. 2014; Florence. Florence: IEEE; 2014. p. 1-7. DOI: 10.1109/IEVC.2014.7056090

[13] Ramteen S, Denholm P. Emissions impacts and benefits of plug-in hybrid electric vehicles and vehicle-to-grid services emissions impacts and benefits of plug-in hybrid electric vehicles and vehicle-to-grid. Environmental Science & Technology. 2009;43(4): 1199-1204. DOI: 10.1021/es802324j

[14] Samaras C, Meisterling K. Life cycle assessment of greenhouse gas emissions from plug-in hybrid vehicles: implications for policy. Environmental Science & Technology. 2008;42(9):3170-3176. DOI: 10.1021/es702178s

[15] Guille C, Gross G. A conceptual framework for the vehicle-to-grid (V2G) implementation. Energy Policy. 2009;37(11):4379-4390. DOI: 10.1016/j.enpol.2009.05.053

[16] Tomić J, Kempton W. Using fleets of electric-drive vehicles for grid support. Journal of Power Sources. 2007;168(2):459-468. DOI: 10.1016/j.jpowsour.2007.03.010

[17] De Los Ríos A, Goentzel J, Nordstrom K, Siegert C. Economic analysis of vehicle-to-grid (V2G)-enabled fleets participating in the regulation service market. In: IEEE, editor. 2012 IEEE PES Innovative Smart Grid Technologies (ISGT); 16–20 Jan. 2012; Washington, DC. Washington, DC: IEEE; 2012. p. 1-8. DOI: 10.1109/ISGT.2012.6175658

[18] Wu D, Chau K, Liu C, Gao S, Li F. Transient stability analysis of SMES for smart grid with vehicle-to-grid operation. IEEE Transactions on Applied Superconductivity. 2012;22(3):1-5. DOI: 10.1109/TASC.2011.2174572

[19] Wang Z, Wang S. Grid power peak shaving and valley filling using vehicle-to-grid systems. IEEE Transactions on Power Delivery. 2013;28(3):1822-1829. DOI: 10.1109/TPWRD.2013.2264497

[20] Rautiainen A, Markkula J, Repo S, Kulmala A, Jarventauta P, Vuorilehto K. Plug-in vehicle ancillary services for a distribution network. In: IEEE, editor. 2013 8th Interna-

tional Conference and Exhibition on Ecological Vehicles and Renewable Energies; 27–30 March 2013; Monte Carlo. Monte Carlo: IEEE; 2013. p. 1-8. DOI: 10.1109/EVER.2013.6521524

[21] Sortomme E, El-Sharkawi M. Optimal scheduling of vehicle-to-grid energy and ancillary services. IEEE Transactions on Smart Grid. 2012;3(1):351-359. DOI: 10.1109/TSG.2011.2164099

[22] Ilić D, Karnouskos S, Beigl M. Improving accuracy of energy forecasting through the presence of an electric vehicle fleet. Electric Power Systems Research. 2015;120(1):32-38. DOI: 10.1016/j.epsr.2014.10.008

[23] Vithayasrichareon P, Mills G, MacGill I. Impact of electric vehicles and solar PV on future generation portfolio investment. IEEE Transactions on Sustainable Energy. 2015;6(3):899-908. DOI: 10.1109/TSTE.2015.2418338

[24] Denholm P, Kuss M, Margolis R. Co-benefits of large scale plug-in hybrid electric vehicle and solar PV deployment. Journal of Power Sources. 2013;236(1):350-356. DOI: 10.1016/j.jpowsour.2012.10.007

[25] Yilmaz M, Krein P. Review of the impact of vehicle-to-grid technologies on distribution systems and utility interfaces. IEEE Transactions on Power Electronics. 2013;28(12):5673-5689. DOI: 10.1109/TPEL.2012.2227500

[26] Clement-Nyns K, Haesen E, Driesen J. The impact of charging plug-in hybrid electric vehicles on a residential distribution grid. IEEE Transactions on Power Systems. 2010;25(1):371-380. DOI: 10.1109/TPWRS.2009.2036481

[27] Esmaili M, Rajabi M. Optimal charging of plug-in electric vehicles observing power grid constraints. IET Generation, Transmission & Distribution. 2014;8(4):583-590. DOI: 10.1049/iet-gtd.2013.0628

[28] Hajforoosh S, Masoum M, Islam S. Real-time charging coordination of plug-in electric vehicles based on hybrid fuzzy discrete particle swarm optimization. Electric Power Systems Research. 2015;128(1):19-29. DOI: 10.1016/j.epsr.2015.06.019

[29] Han S, Han S, Sezaki K. Development of an optimal vehicle-to-grid aggregator for frequency regulation. IEEE Transactions on Smart Grid. 2010;1(1):65-72. DOI: 10.1109/TSG.2010.2045163

[30] Sortomme E, El-Sharkawi M. Optimal charging strategies for unidirectional vehicle-to-grid. IEEE Transactions on Smart Grid. 2011;2(1):119-126. DOI: 10.1109/TSG.2010.2090910

[31] Ansari M, Al-Awami A, Sortomme E, Abido M. Coordinated bidding of ancillary services for vehicle-to-grid using fuzzy optimization. IEEE Transactions on Smart Grid. 2015;6(1):261-270. DOI: 10.1109/TSG.2014.2341625

[32] Qian K, Zhou C, Allan M, Yuan Y. Modeling of load demand due to EV battery charging in distribution systems. IEEE Transactions on Power Systems. 2011;26(2):802-810. DOI: 10.1109/TPWRS.2010.2057456

[33] Papadopoulos P, Jenkins N, Cipcigan L, Grau I, Zabala E. Coordination of the charging of electric vehicles using a multi-agent system. IEEE Transactions on Smart Grid. 2013;4(4):1802-1809. DOI: 10.1109/TSG.2013.2274391

[34] Ahmad M, Musirin I, Othman M. Optimal charging strategy for plug-in hybrid electric vehicle using evolutionary algorithm. In: IEEE, editor. 2014 IEEE 8th International Power Engineering and Optimization Conference (PEOCO2014); 24–25 March 2014; Langkawi. Langkawi: IEEE; 2014. p. 557-562. DOI: 10.1109/PEOCO.2014.6814491

[35] Geng B, Mills J, Sun D. Two-stage charging strategy for plug-in electric vehicles at the residential transformer level. IEEE Transactions on Smart Grid. 2013;4(3):1442-1452. DOI: 10.1109/TSG.2013.2246198

[36] Oliveira D, De Souza A, Delboni L. Optimal plug-in hybrid electric vehicles recharge in distribution power systems. Electric Power Systems Research. 2013;98:77-85. DOI: 10.1016/j.epsr.2012.12.012

[37] Soares F, Almeida P, Lopes J. Quasi-real-time management of electric vehicles charging. Electric Power Systems Research. 2014;108:293-303. DOI: 10.1016/j.epsr.2013.11.019

[38] Hu J, You S, Lind M, Østergaard J. Coordinated charging of electric vehicles for congestion prevention in the distribution grid. IEEE Transactions on Smart Grid. 2014;5(2):703-711. DOI: 10.1109/TSG.2013.2279007

[39] Gan L, Topcu U, Low S. Optimal decentralized protocol for electric vehicle charging. In: IEEE, editor. 2011 50th IEEE Conference on Decision and Control and European Control Conference (CDC-ECC); 12-15 December 2011; Orlando, USA. Orlando, USA: IEEE; 2013. p. 5798-5804. DOI: 10.1109/TPWRS.2012.2210288

[40] Vandael S, Claessens B, Hommelberg M, Holvoet T, Deconinck G. A scalable three-step approach for demand side management of plug-in hybrid vehicles. IEEE Transactions on Smart Grid. 2013;4(2):720-728. DOI: 10.1109/TSG.2012.2213847

[41] Weihao H, Su C, Chen Z, Bak-Jensen B. Optimal operation of plug-in electric vehicles in power systems with high wind power penetrations. IEEE Transactions on Sustainable Energy. 2013;4(3):577-585. DOI: 10.1109/TSTE.2012.2229304

[42] Bai X, Qiao W. Robust optimization for bidirectional dispatch coordination of large-scale V2G. IEEE Transactions on Smart Grid. 2015;6(4):1944-1954. DOI: 10.1109/TSG.2015.2396065

[43] Kavousi-Fard A, Rostami A, Niknam T. Reliability-oriented reconfiguration of vehicle-to-grid networks. IEEE Transactions on Industrial Informatics. 2015;11(3):682-691. DOI: 10.1109/TII.2015.2423093

[44] Ghofrani M, Arabali A, Ghayekhloo M. Optimal charging/discharging of grid-enabled electric vehicles for predictability enhancement of PV generation. Electric Power Systems Research. 2014;117:134-142. DOI: 10.1016/j.epsr.2014.08.007

[45] Shaaban F, Ismail M, El-Saadany E, Zhuang W. Real-time PEV charging/discharging coordination in smart distribution systems. IEEE Transactions on Smart Grid. 2014;5(4): 1797-1807. DOI: 10.1109/TSG.2014.2311457

[46] Haddadian G, Khalili N, Khodayar M, Shahidehpour M. Optimal scheduling of distributed battery storage for enhancing the security and the economics of electric power systems with emission constraints. Electric Power Systems Research. 2015;124:152-159. DOI: 10.1016/j.epsr.2015.03.002

[47] Nguyen H, Zhang C, Mahmud A. Optimal coordination of G2V and V2G to support power grids with high penetration of renewable energy. IEEE Transactions on Transportation Electrification. 2015;1(2):188-195. DOI: 10.1109/TTE.2015.2430288

[48] Singh M, Thirugnanam K, Kumar P, Kar I. Real-time coordination of electric vehicles to support the grid at the distribution substation level. IEEE Systems Journal. 2015;9(3): 1000-1010. DOI: 10.1109/JSYST.2013.2280821

[49] Liu H, Hu Z, Song Y, Wang J, Xie X. Vehicle-to-grid control for supplementary frequency regulation considering charging demands. IEEE Transactions on Power Systems. 2015;30(6):3110-3119. DOI: 10.1109/TPWRS.2014.2382979

11

Aging and Degradation Behavior Elucidated by Viscoelasticity Aiming Protection of Smart City Facilities

Yukitoshi Takeshita, Takashi Miwa, Azusa Ishii and Takashi Sawada

Abstract

Polymer coatings play a crucially important role in protecting smart city facilities against the harsh factors of outdoor environments. Recent increased awareness of eco-friendliness has led to the use of waterborne organic coatings. Research into the bulk material properties of these coatings is necessary in order to understand their degradation process in the field. The present work focuses attention on a unique rheological property, which has both elastic and viscous characteristics, as a means of assessing the stability of the coating. The viscoelastic property determines whether it presents solid-like or liquid-like response from the comparison of relative strengths of the relaxation time (τ) and operating time (t). In the process of degradation, both the storage (E′) and loss modulus (E″), which represent the elastic and viscous components, respectively, decrease accordingly, reflecting the deterioration of coating. The majority of the water molecules absorbed in a coating are strongly bound to the polymer network through hydrogen bonds with polar functional groups, which destroys intermolecular bonding between macromolecules and reduces the bulk materials' ability to diffuse stress concentrations and thereby lowers a coating's overall strength.

Keywords: rheology, static and dynamic viscoelasticity, aging and degradation, water diffusion and absorption, FT-IR, DSC

1. Introduction

In providing telecommunications services throughout Japan, NTT utilizes enormous numbers of telecommunications plants and materials, including approximately two-million kilometers of cable and nearly 12 million telephone poles [1]. Outdoor materials in particular are exposed to a wide range of environments: UV light, a range of temperatures and humidity, sea salt particles, alkalinity, and acidity. Therefore, these materials are protected by some type of weathering-prevention technology, and a wide variety of anti-corrosion technologies have been developed for each type of environment [2].

Polymeric materials are available for coating onto large steel structures, such as road and railway bridges, and electric power and telephone poles, which form a part of the social infrastructure in smart cities. We have therefore developed and introduced organic coatings for use in industrial telecommunications facilities [3–7].

The coatings typically have a multilayer structure, with the bottom layers functioning to bond the coating to the steel substrate and the top layers providing primary protection and pigmentation. Conventional coatings usually include a wide variety of volatile organic compounds (VOC), such as hydrocarbons (hexane, toluene, xylene), ketones (acetone, methyl ethyl ketone, methyl isobutyl ketone), alcohols (methanol, ethanol, cyclohexanol), and esters (ethyl acetate, butyl acetate, isobutyl acetate) [8]. VOCs are environmentally damaging and pose human health risks. So, an awareness of the importance of being eco-friendly has been growing even in the coating fields.

One solution is to reduce the proportion of solvent in the coating, ultimately to zero, and we have already developed and introduced a low-solvent coating for use in the telecommunications field. Another solution is to replace organic solvents with water. Water-based coatings have been in high demand, and we have therefore been studying them with a view to protecting human health and the environment [3–7, 9–11]. Research into the bulk material properties of these low-VOC coatings is necessary to understand their degradation process.

In engineering applications of polymeric materials, physical aging commonly arises throughout the service lifetime of the component. Physical aging is proposed to be a long-term relaxation process that occurs in polymers in a glassy state below the glass transition temperature (T_g) as the macromolecules gradually approach thermodynamic equilibrium [12]. This aging is the main phenomenon affecting the physical stability of polymeric materials and thus of organic coatings, and it is accepted that this phenomenon has an impact on the formulation, application, and service life of organic coatings. While the studies reported to date clearly indicate the considerable practical importance of the aging phenomenon in relation to degradation behavior and failure, theoretical and practical questions remain unsolved [12, 13].

The present work focuses on the effect of aging and further degradation on the rheological property of mainly epoxy/urethane coating film, which is targeted for use in telecommunications fields. The static and dynamic viscoelastic properties are intensively evaluated in an accelerated weathering condition, and the correlation between degradation behaviors and rheological properties is discussed. Finally, the states of water absorbed in coating as an aggressive factor affecting property of coating are represented using differential scanning calorimetry (DSC) and Fourier transform infrared (FTIR) spectroscopy.

2. Past research

Physical aging is an important factor in relation to the physical properties of many polymers at room temperature, which is below T_g [14–17]. A lot of past researches have investigated the physical aging of polymers including polyethylene [18, 19], epoxy [20–22], poly(methyl methacrylate) [23, 24], polyester [25, 26], poly(vinyl acetate) [27], powder coatings [28, 29], and organic coatings [12, 30–35]. There has been a great deal of excellent work on the spontaneous relaxation process of coating films during physical aging [12, 14, 18–20, 28–31]. Skaja has published outstanding basic work designed to determine the relationship between changes in bulk mechanical properties and degradation during the accelerated weathering of coating films [31].

In research on adhesives, it has been well established through many studies that the performance of pressure-sensitive adhesives (e.g., peel, tack, and shear) depends strongly on the bulk viscoelastic properties of the adhesives [36–45]. Further, as regards wood research, Mizumachi et al. have suggested that adhesive strength is strongly dependent on the rheological properties of the adhesives [46–54]. However, the exact relationship between rheological performance and degradation behavior for eco-friendly coating film has not been established.

3. Theory

3.1. Physical aging

The basic thermodynamic description of the state of the glass for polymers has been universally accepted for many years. Indeed, the earliest thermodynamic treatments by Davies and Jones [55, 56] remain among the best available. Perera has reported theoretical considerations on physical aging of polymer [12].

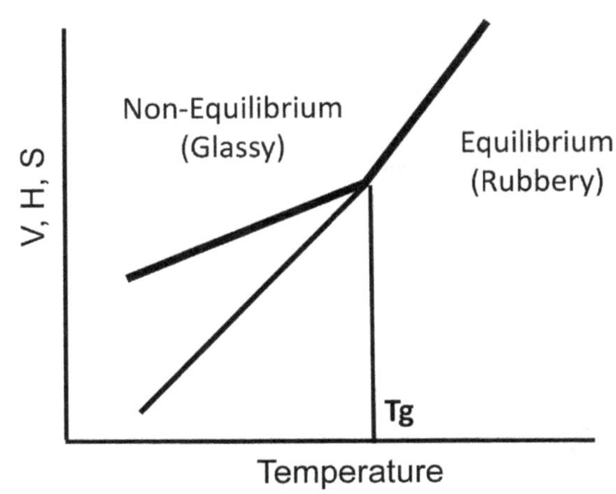

Figure 1. Schematic representation of the dependence of volume (V), enthalpy (H) and entropy (S) on temperature, based on Perera [12].

Polymer volume (V), enthalpy (H), and entropy (S) are represented as a function of temperature (T) as shown in **Figure 1**. During cooling, V, H, and S behave differently at temperatures above and below T_g. At $T>T_g$, the decrease in V, H, and S can follow the decrease in temperature due to higher mobility of macromolecular chains, and thus the polymer is in equilibrium state. At $T<T_g$, due to lower mobility of macromolecular chains, the decrease in V, H, and S cannot follow the decrease in temperature and thus the polymer is in a nonequilibrium state. In this state, molecular motion is limited but does not stop and continues toward an equilibrium state. This spontaneous process has been known under many terms, such as volume relaxation, enthalpy relaxation, mechanical relaxation, and structural relaxation. These relaxations are referred as "physical aging" [12].

3.2. Rheology

Rheology is the study addressing the deformation and the flow of materials [57]. It applies to substances with a complex microstructure, such as muds, sludges, suspensions, polymers, and other glass formers, as well as to many foods and additives, bodily fluids, and other biological materials, or to other materials in the class of soft matter. The rheological response is expressed as the combination of elastic and viscous components which has elastic modulus and viscosity. It is normally represented using Maxwell model consisting of a spring with modulus E in series with a dashpot with viscosity η [**Figure 2(a)**] and Voigt model consisting of a spring in parallel with a dashpot [**Figure 2(b)**].

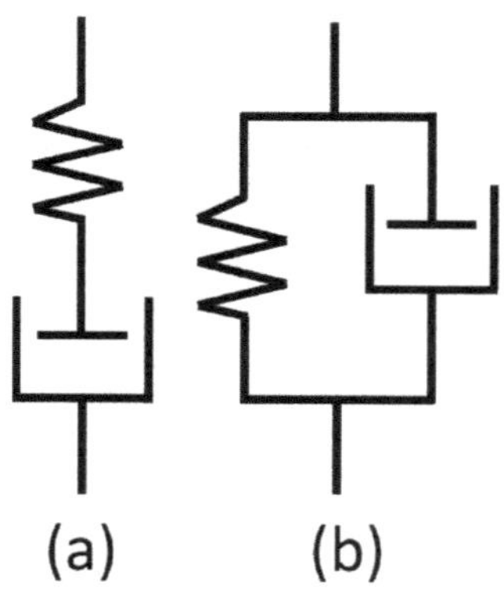

Figure 2. Diagram of viscoelastic material. (a) Maxwell model and (b) Voigt model.

3.2.1. Static viscoelasticity

3.2.1.1. Maxwell model

When constant strain ε is applied to Maxwell model, stress σ in each component is expressed as:

$$\sigma = E\,\varepsilon_1 \tag{1}$$

$$\sigma = \eta \frac{d\varepsilon_2}{dt} \tag{2}$$

where

$$\varepsilon = \varepsilon_1 + \varepsilon_2 \tag{3}$$

E and η are elastic modulus and viscosity. By solving simultaneous Eqs. (1)–(3),

$$\sigma = \sigma_0 e^{-t/\tau} \tag{4}$$

where σ_0 is an initial stress, t is time, and τ is relaxation time, η/E.

3.2.1.2. Voigt model

When constant stress σ is applied to Voigt model, stress in each component is expressed as:

$$\sigma_1 = E\,\varepsilon \tag{5}$$

$$\sigma_2 = \eta \frac{d\varepsilon}{dt} \tag{6}$$

where

$$\sigma = \sigma_1 + \sigma_2 \tag{7}$$

By solving simultaneous Eqs (5)–(7),

$$\varepsilon = \frac{\sigma_0}{E}\left\{1 - e^{-t/\tau}\right\} \tag{8}$$

3.2.2. Dynamic viscoelasticity

When sinusoidal strain ε expressed as:

$$\varepsilon = \varepsilon_0 e^{i\omega t} \tag{9}$$

is given to Maxwell model,

$$\sigma = \sigma_0 e^{i(\omega t+\delta)} \tag{10}$$

where $i = \sqrt{-1}$, ω and δ are the angular velocity and phase difference. By solving simultaneous Eqs (1)–(3), (9) and (10), complex modulus E* is expressed as:

$$E^* = \sigma / \varepsilon = \left(\frac{\omega^2\tau^2}{1+\omega^2\tau^2} + i\frac{\omega\tau}{1+\omega^2\tau^2} \right) E = E' + iE'' \tag{11}$$

where E′ is the storage modulus $\left(= \frac{\omega^2\tau^2}{1+\omega^2\tau^2} E \right)$ and E″ is the loss modulus $\left(= \frac{\omega\tau}{1+\omega^2\tau^2} E \right)$. **Figure 3** shows E′/E and E″/E vs ωτ. At low frequency, the model behaves more viscously, while at high frequency, it behaves more elastically. In this work, stress σ in Eq. (4), storage modulus E′ and loss modulus E″ in Eq. (11), and loss tangent tanδ (E″/E′) are essentially discussed.

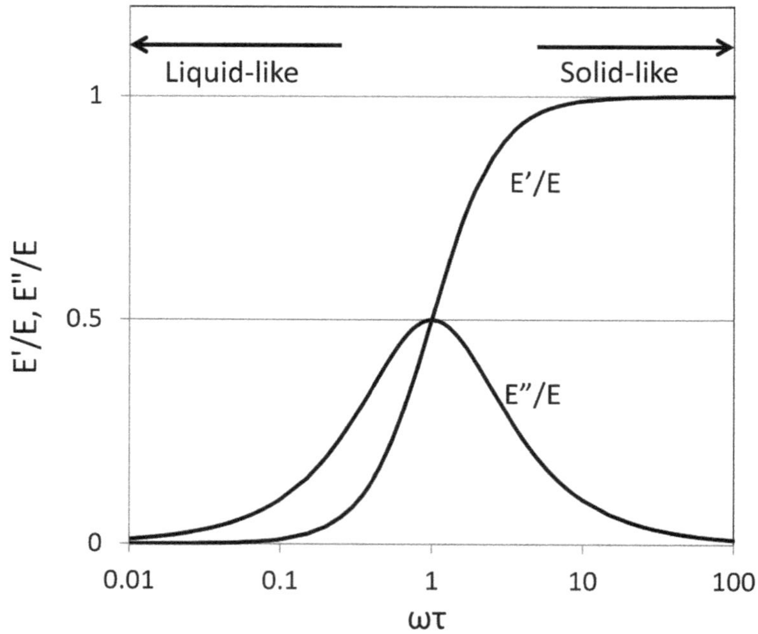

Figure 3. E′/E, E″/E vs. ωτ for Maxwell model.

4. Dynamic viscoelasticity in physical aging

To understand the basics of the behaviors of coating aged at room temperature, the dynamic viscoelastic properties, E′ and tanδ, were determined using a Rheolograph Solid (S-1, Toyo-seiki, Tokyo, Japan) [5]. The initial tension and dynamic strain are 60 gf (0.588N) and 8 μm, respectively. Coating films A, B, and C are all eco-friendly waterborne coatings. Each comprises a bottom, middle, and top layer. The structures are respectively epoxy/epoxy/inorganic with total thickness of 130 μm for A, epoxy/epoxy/urethane with 125 μm for B, and acryl/acryl-styrene/acryl-urethane with 275 μm for C. **Figure 4(a)** and **(b)** shows the effect of aging on E′ and tanδ for coating film from 3 to 48 days old. Only the results for 100 Hz are shown for a discussion of the overall trend, though measurements for four frequencies, 100, 10, 1, and 0.1 Hz, were performed.

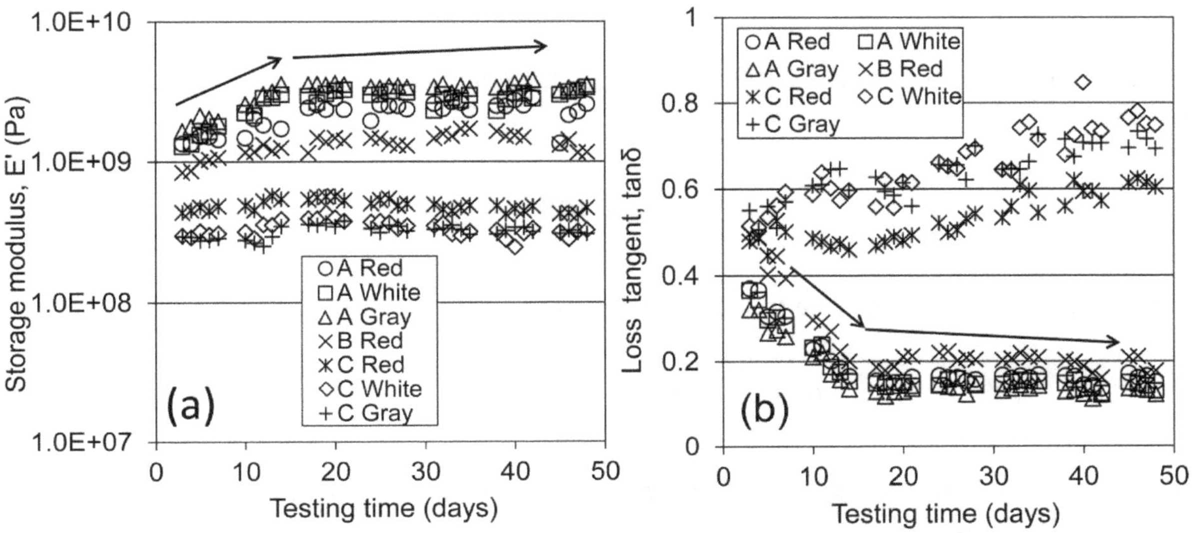

Figure 4. Effect of aging on E′ and tanδ for coating film aged 3–48 days at 100 Hz. (a) E′ and (b) tan δ [5].

From **Figure 4(a)**, it can be seen that the viscoelastic properties of A and B change rapidly for the first few days after preparation and then stabilize after a few weeks. After this, the sample no longer changes much with age. Although the data are not presented here, lower frequencies experience greater changes with increasing sample age, though stability occurs at the same age. **Figure 4(b)** shows that tanδ for A and B decreases with age, though it does so at low frequencies equally.

It is considered that the viscoelastic properties of A and B change because of the evaporation of the solvent that remained in the coating matrix. As the solvent evaporates, the film becomes hard and rigid and exhibits a high E′. In addition, the densification and the cross-link reaction might proceed [12]. As a result, the elastic property increased, corresponding to the increase in E′, and the viscous property decreased, corresponding to the decrease in tanδ. Once the solvent has completely evaporated, it might no longer affect the viscoelastic properties and become stable.

Film C exhibits very different trends. From **Figure 4(a)**, its E′ value seems to change very little with age. In addition, tanδ increases with age as shown in **Figure 4(b)**. We believe that C fundamentally has a strong viscous feature compared with A and B. Therefore, even after considerable time has elapsed, it is only slightly more rigid than at the beginning. As described in the previous section, E′ generally represents the elastic feature and E″, thus tanδ, indicates the viscous one. At a simple level, it can determine the characteristics representing the nature of a film which has solidity or softness from dynamic viscoelastic measurements.

As coatings A and B exhibit the elastic property, the relaxation time τ is considered here. From **Figure 3**, the region that represents the elastic property is the right area where x values are around 5–100:

$$5 < \omega\tau < 100 \tag{12}$$

The frequency f applied here is 100 Hz, so angular velocity ω is obtained by the following equation.

$$\omega = 2\pi f \fallingdotseq 628 \tag{13}$$

From Eqs. (12) and (13), the relaxation time τ is expressed as follows:

$$0.008\ (\fallingdotseq 0.01) < \tau < 0.16 \tag{14}$$

Operating time t is the reciprocal of the frequency, which is 0.01(=1/100) (s). Therefore, the following relation is introduced.

$$\left[\mathit{Relaxation\ time}\ (\tau)\right] > \left[\mathit{Operating\ time}\ (t)\right]$$

From this relation that relaxation time is higher than operating time, it is understood that the material does not flow (not relax) and thus presents a solid-like response and behaves as an elastic body.

Similarly, for coating C, the region that represents the viscous property is the left area where x values are around 0.01–0.5:

$$0.01 < \omega\tau < 0.5 \tag{15}$$

From Eqs. (13) and (15),

$$0.000016 < \tau < 0.0008 \tag{16}$$

As operating time t is 0.01 (=1/100) (s), the following relation is introduced.

$$\left[\mathit{Relaxation\ time}\ (\tau)\right] < \left[\mathit{Operating\ time}\ (t)\right]$$

That is, relaxation time is lower than operating time. This causes the material to flow (relax) and thus present a liquid-like response and behave as a viscous body.

Figure 5 illustrates the viscoelastic properties with the failure modes. At high E′, the breaking location is the bulk region of the layer. For high tanδ, it is the interface of the two layers. It is considered that E′ corresponds to the rigidity of the bulk material and the inability to release another material at an interface; tanδ is related to both the ability of the material to adjust to suit stress and to grasp another layer at the interface. Therefore, a high E′ indicates a high likelihood of failure within bulk planes, and a high tanδ increases the likelihood of failure within interface planes. A high contrast between the E′ values of sequential layers (high E′ and low E′) also increases the interface failure rate.

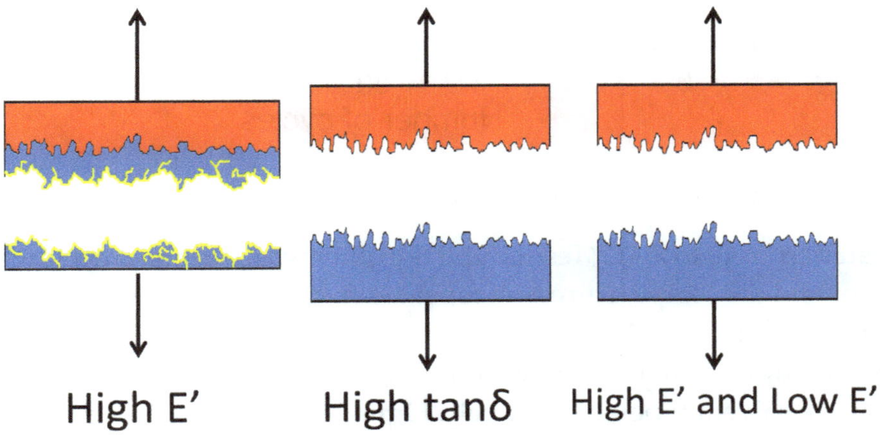

Figure 5. Viscoelastic properties and failure mode.

5. Behavior in heat cycling test

Figure 6 shows viscoelastic values of E′ and E″ of waterborne epoxy/acryl-urethane with 100-μm thickness under a heat cycling test that repeats a cycle from -30 to 70°C [6], where one cycle takes 12 h. Relative humidity reaches 90% at high temperature of 70°C and is almost 0% below 0°C. **Figure 6** shows that E′ increases at 50 cycles and then decreases at 100 cycles. In particular, several small cracks are observed on the coating at 100 cycles. In addition, T_g increases at around 50°C at 50 cycles; however, it does not vary from 50 to 100 cycles. Taking the results in **Figure 4** into account, the region of 0–50 cycles exhibits an increase in E′ and decrease in E″, similar to the aging behavior observed in coatings A and B in **Figure 4**; however, the region of 50–100 cycles shows dissimilar degradation behavior, where both E′ and E″ decrease due to water absorption.

It is found that the heat cycling accelerates degradation more than room-temperature aging does. To determine the effect of dry and wet states under the heat cycling condition, the degradation behavior under constant humidity, dry or wet, was investigated [7]. Two tests were performed: one was a dry heat cycle where humidity was 10% constant and the temper-

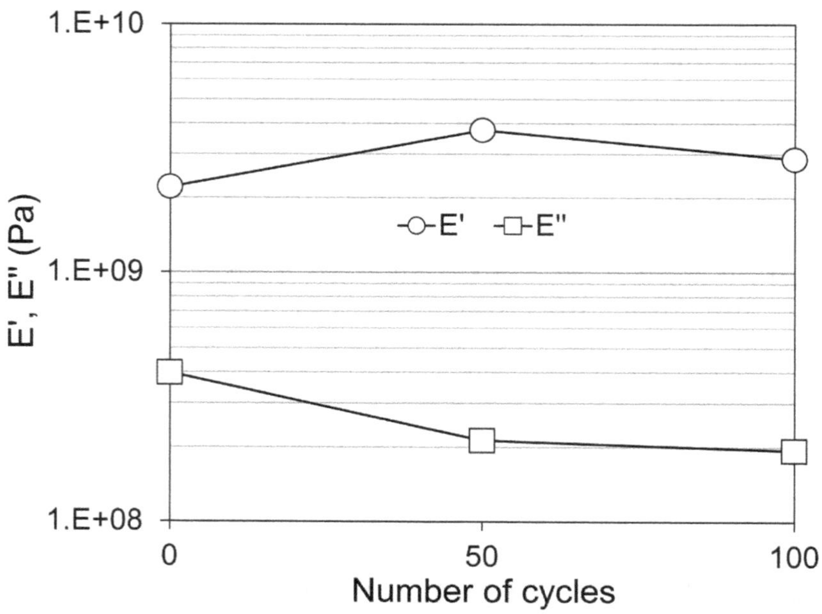

Figure 6. Effect of heat cycling test on E′ and E″ for epoxy/acryl-urethane [6].

ature was repeated from 10 to 70°C (Test 1), and the other was a wet heat cycle where humidity was 95% constant and the temperature was repeated from 10 to 70°C in the same way (Test 2). E′ and tanδ values are shown in **Figure 7**. The coatings are all epoxy/epoxy/urethane. A and B are waterborne coatings where the epoxy equivalent ratio is smaller for A than for B. C and D are strong and weak solvent types, respectively. **Figure 7** shows that E′ once increases and then stabilizes with age, similar to its behavior in **Figure 4**. Accordingly, tanδ rapidly decreases and then stabilizes. These behaviors are similar to those in coatings A and B in **Figure 4** and in the first half period (0–50 cycles) in **Figure 6**. Strict comparison elucidates that the degrees of increase in E′ and decrease in tanδ are larger in the wet state than in the dry one. Therefore, a comparison of T_g values is shown in **Figure 8**. The peak temperature indicates T_g. Tests 1 and 2 were dry and wet states, respectively. The increase in T_g is obviously larger in the wet state than in the dry one [(a) wet 30°C > dry 20°C; (b) wet 35°C > dry 30°C; (c) wet 45°C > dry 25°C; (d) wet 15°C > dry 10°C]. Thus, the increase in E′ is attributed to curing and hardening of the coating, which causes high T_g that are facilitated remarkably in the wet state more than in the dry state. Densification also accounts for the high T_g. This densification needs high mobility of macromolecule segments. It is also considered that the wet state helps macromolecule movement.

To determine experimentally whether densification proceeds, DSC was performed. **Figure 9** shows the thermal analysis in DSC. Densification can be detected by enthalpy relaxation that shows an endothermic peak due to absorbing the heat in the sub-T_g region. Black arrows certainly indicate the enthalpy relaxation in waterborne coatings A and B in **Figure 9**. However, solvent-type coatings C and D do not exhibit it. Instead, an endothermic peak due to evaporation of the solvent is observed as indicated by the white arrow. Therefore, for waterborne coating, high T_g is partly due to densification, whereas, for solvent-type coating, it is somewhat due to solvent evaporation.

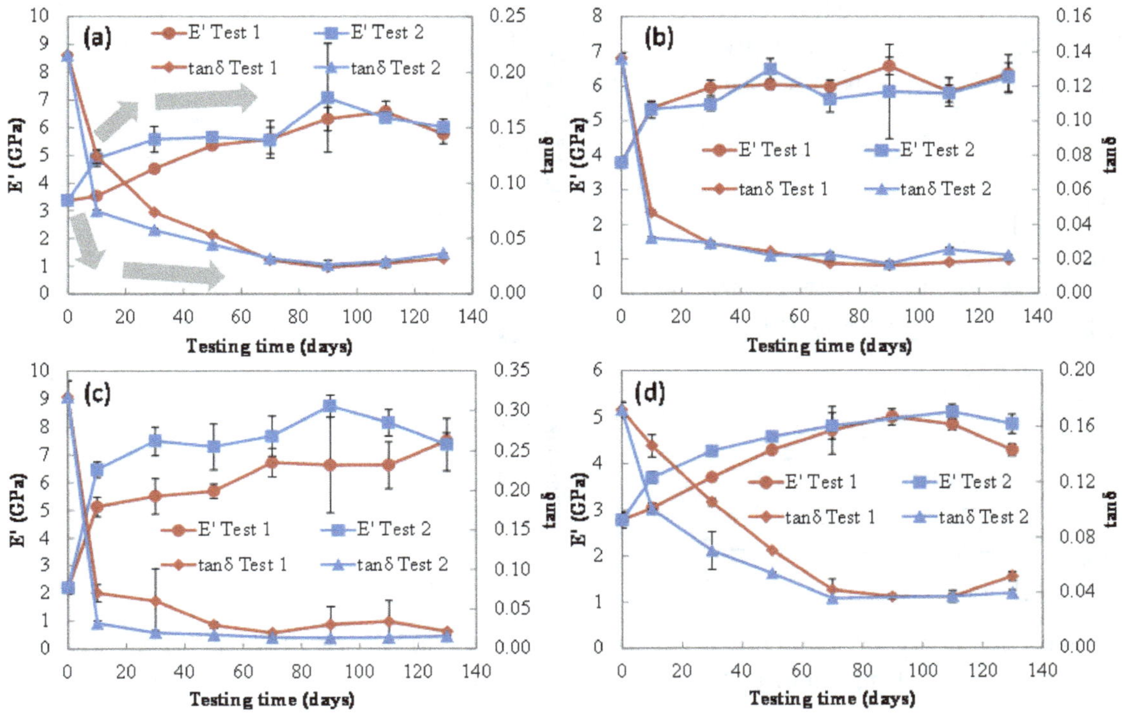

Figure 7. E′ and tan δ in heat cycling test. Test 1: dry. Test 2: wet. (a) Coating A, (b) coating B, (c) coating C, (d) coating D [7].

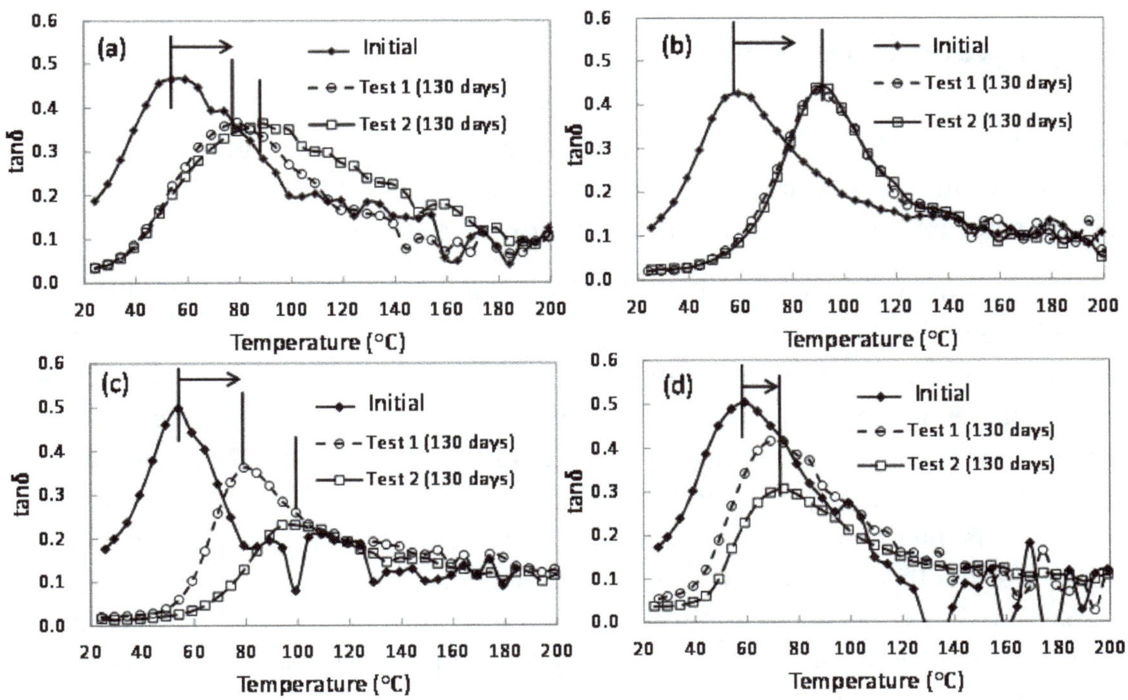

Figure 8. Temperature dependence of tan δ at 130 days in heat cycling test. Test 1: dry. Test 2: wet. (a) Coating A, (b) coating B, (c) coating C, (d) coating D [7].

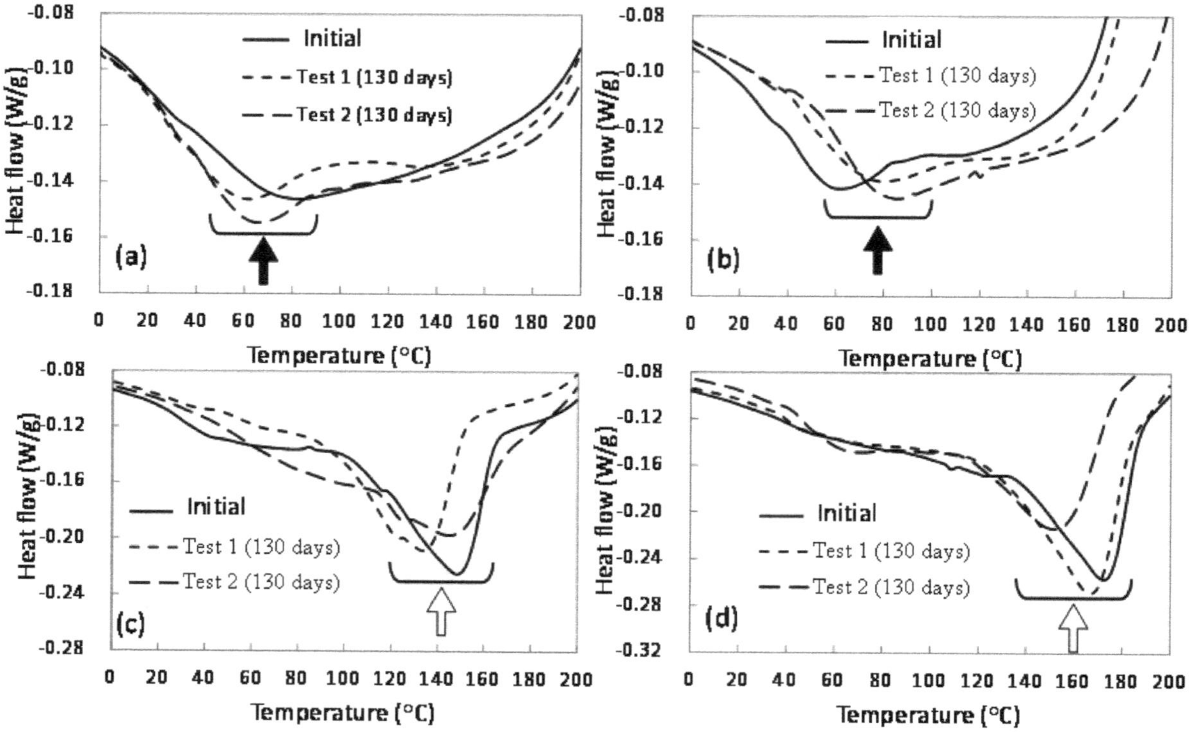

Figure 9. DSC curves at 130 days in heat cycling test. Test 1: dry. Test 2: wet. (a) Coating A, (b) coating B, (c) coating C, (d) coating D [7].

6. Effect of water molecules

As it is understood that the high-moisture state accelerates the aging and degradation behaviors, the influence of water molecules is focused on in this section [9]. The water diffusion process and saturated water content, as well as adhesive strength, are investigated.

Figure 10 shows the nature of water diffusing in coating film. Coatings A and B are epoxy/epoxy/urethane with 125-μm thickness and epoxy/epoxy/fluororesin with 140-μm thickness, respectively. Both are waterborne coatings. From **Figure 10**, water content M_t increases with time and stabilizes around 50–100 h. The saturated contents M_∞ of A and B are 6.6% and 3.7%, respectively. The higher M_∞ of coating A is attributed to higher water absorption of urethane than that of the fluororesin of the top layer. Bottom layers are both epoxy, but the polarity of the epoxy of A is likely higher than that of B. **Figure 11** shows the aging time dependence of M_∞. The longer the time is, the lower the M_∞ value becomes (100 days: A 8.5%, B 5.5% > 200 days: A 6.6%, B 3.7%). The decrease in M_∞ is caused by the densification of macromolecules. This densification process progresses over aging time due to the relaxation of macromolecule segments that occurs in the coatings glassy state [12]. As the densification progresses, the β transition peak shifts to a higher temperature, which is a phenomenon that has been observed in our research. This agrees with the above aging time dependence of M_∞ in **Figure 11**.

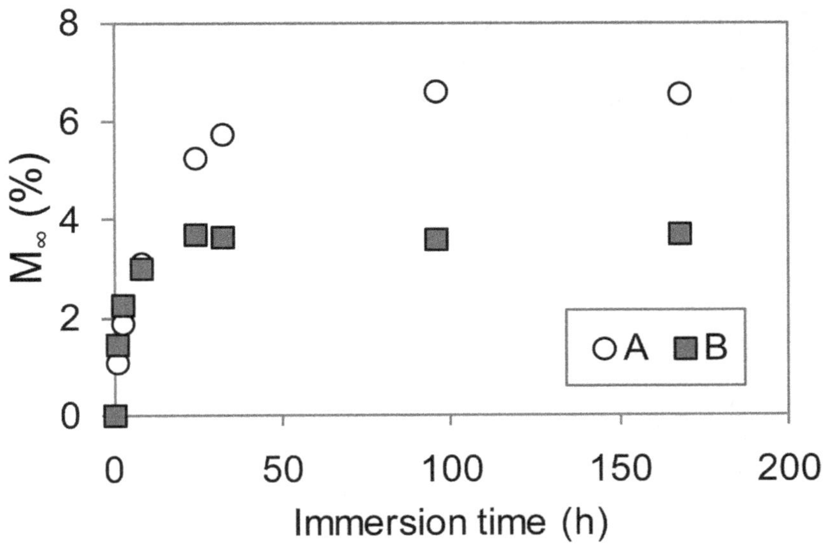

Figure 10. Water absorption behavior. A epoxy/epoxy/urethane. B epoxy/epoxy/fluororesin [9].

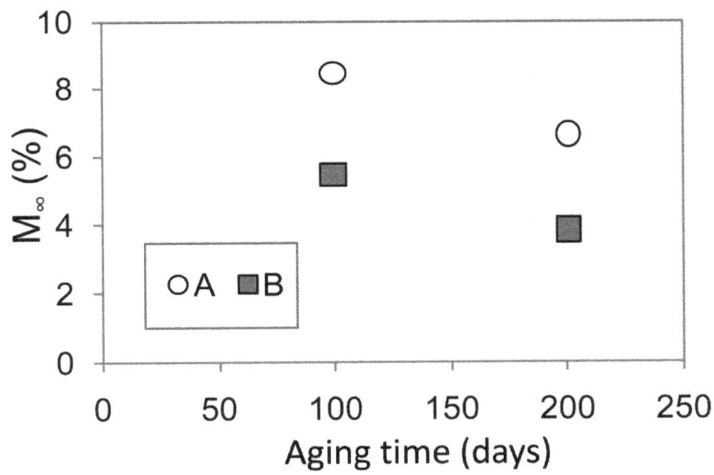

Figure 11. Aging time dependence of saturated water content. A epoxy/epoxy/urethane. B epoxy/epoxy/fluororesin [9].

The dynamics of water diffusion is studied next. The M_t/M_∞ linearly increases at the early stage of absorption and steadily approaches equilibrium. This feature indicates that water absorption exhibits Fickian diffusion [58] and the diffusion coefficient D can be calculated from the slope of plot of M_t/M_∞ vs $t^{1/2}l^{-1}$, where l is film thickness. D values of A and B are 7.39×10^{-14} and 24.2×10^{-14} m^2/s, that is, A < B, which is opposite to the saturated water content, M_∞. The difference in D seems to be due to the difference in the cross-link density. By measurement of E′ in the plateau region of E′ vs temperature, it is found that the cross-link densities are around 5300 and 2600 mol/m^3 for A and B. That is, the lower D shows the higher crosslink density. This result suggests that higher crosslink density prevents water molecules from easily diffusing.

In relation to the mechanical properties in water absorption, **Figure 12** shows adhesive strengths, which agree with the trends in the E′ values. Adhesive strengths once increase at 50 cycles and then decrease at 100 cycles. At 100 cycles, they are 0.8 and 3.7 MPa for A and B, which is A < B. Taking the results of A > B for M_∞ and A < B for *D* into account, it is clear that the M_∞ value affects the degraded adhesive performance much more than the *D* value does. This might be because the water content is saturated for most of the moisture period in the test.

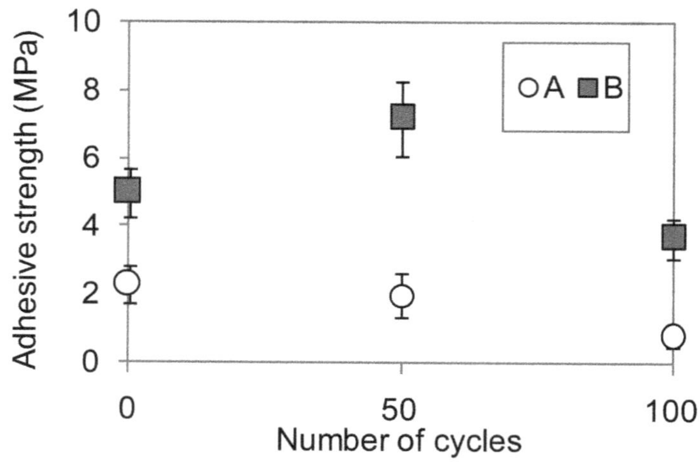

Figure 12. Adhesive strength in heat cycling test. A epoxy/epoxy/urethane. B epoxy/epoxy/fluororesin [9].

7. Static viscoelasticity during water absorption

As it is understood that the saturated water significantly affects the deterioration of E′ and adhesive strength, we focus attention on water molecules as an aggressive factor in this section [10]. That is, we attempt to determine their contribution to static stress relaxation behavior as a mechanical property of coatings.

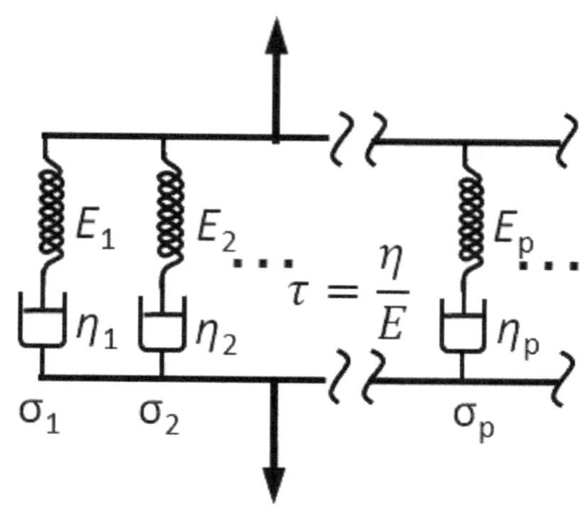

Figure 13. Diagram of generalized Maxwell model for viscoelastic material.

Stress relaxation behavior can be described by a simple Maxwell model of viscoelastic material using an exponential decay function [59]. The stress will decay exponentially, with a relaxation time τ, as described by Eq. (4). Expansion to a generalized Maxwell model, or Maxwell-Weichert model, which consists of several Maxwell model components with different E and η values in parallel, as shown in **Figure 13**, produces a function where the relaxation behavior is described as a superposition of multiple exponentially decaying modes:

$$\sigma(t)/\sigma_0 = \sum_{p=1}^{p\geq 1} g_p\, e^{-(t/\tau_p)} \tag{17}$$

where τ_p is the relaxation time and g_p is the normalized intensity of the pth relaxation mode. The complex nature of the equation typically results in its being simplified into a stretched exponential function, commonly called the William-Watts (W-W) equation [60]:

$$\varphi(t) = \sigma(t)/\sigma_0 = e^{-(t/\tau)^{\beta}} \tag{18}$$

Here, stretching exponent β, with bounds $0< \beta<1$, represents the relative distribution of the relaxation modes; a decrease in β corresponds to an increase in the distribution of relaxation times, τ_p, of the individual components of the system [61–64]. In this equation, τ represents a "characteristic relaxation time" for the system as a whole. The φ(t) is the ratio of the stress at a time "t" to the initial stress.

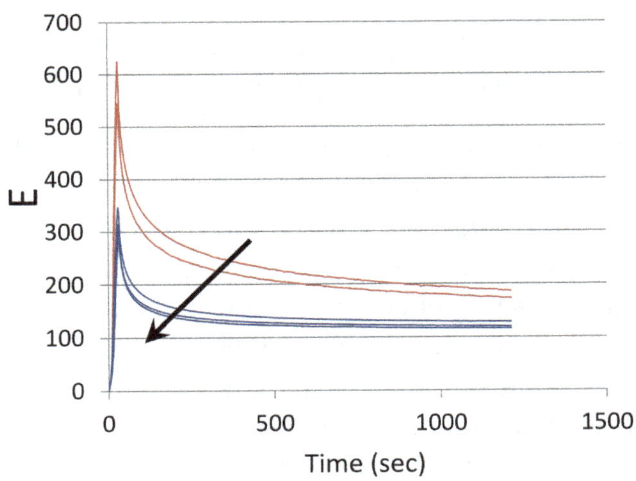

Figure 14. Stress relaxation curves [10].

The three parameters of this equation—σ_0, τ, and β,—are not independent; thus, it is necessary to determine one parameter independently before solving for the other two [15]. In this experiment, σ_0 was taken to be the maximum stress attained by the sample during testing. It may actually be higher as some relaxation occurs as the sample is being drawn; however, this approximation is close enough to provide an accurate description of the behavior of the sample. We measured the changes in the three parameters during water absorption and considered their contribution to the degradation behavior.

The four coatings—A, B, C, and D—are the same as those in section 5. Stress relaxation curves are shown in **Figure 14**. E(t) exponentially decreases with time, which is quite reasonable. With water absorption, the curves shift as indicated by the arrow. The three parameters of the stretched exponential function, E_o, τ, and β, are determined to fit the stress relaxation behavior and plotted as a function of M_t/M_∞ in **Figure 15**. E_o and β quickly decrease in the early stages between $M_t/M_\infty = 0$ and $M_t/M_\infty \approx 0.5$ and level off later between $M_t/M_\infty \approx 0.5$ and $M_t/M_\infty \approx 1$, as shown in **Figure 15(a)** and **(c)**, while τ exhibits a roughly exponential increase between $M_t/M_\infty = 0$ and $M_t/M_\infty \approx 1$, as shown in **Figure 15(b)**.

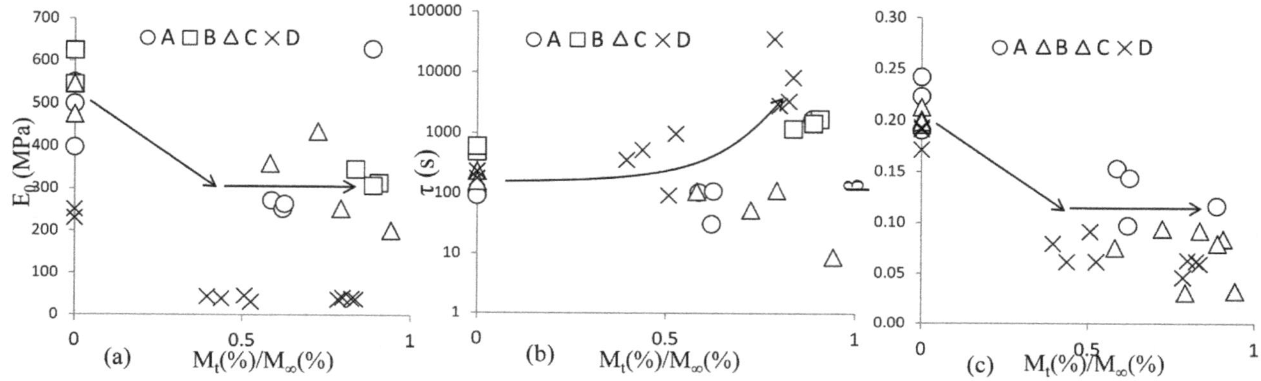

Figure 15. E_o, τ, and β as a function of water content. (a) E_o, (b) τ, (c) β [10].

Water is commonly considered to act as a plasticizing agent in polymers [65–69], which caused a decrease in E_0 in all the samples as M_t/M_∞ approached 1, which is quite reasonable.

Molecular interactions between polymer chains, such as hydrogen and dipole-dipole bonding, cause some degree of reduction in chain mobility [70]. This has the effect of "averaging" or decreasing the distribution of the relaxation times for the individual components of the system. The introduction of water into the polymer network interrupts the intermolecular bonding [66, 71, 72], which reduces this averaging effect and results in the reduction in β for the system as a whole. Water that is bonded to polymer chains interrupts molecular interactions between polymer chains, which contributes to changes in the bulk material properties. Free water, on the other hand, has little effect on polymer chains and little influence on the material properties [71, 73]. The details of behaviors of free water as well as bonded water molecules are described in the next section. As for increase in relaxation time τ in **Figure 15(b)**, absorbed water causes the decrease in restoring force of the elastic component in Maxwell model, which hinders a smooth recovery of the spring component.

The adhesive strength of dry and saturated samples is shown in **Figure 16**. The adhesive strength decreases with water absorption. This indicates that absorbed water has a highly detrimental effect on adhesive strength. As previously asserted, absorbed water interferes with the intermolecular interactions of the polymer. Cross-linking, in addition to narrowing the distribution of relaxation times, may also narrow the distribution of elastic moduli to assist the diffusion of stress concentrations. As described in Eq. (17), each polymer chain, or mode of the system, has its own elastic modulus, which likely follows a distribution similar to the distribution of τ.

Stress applied to the system will concentrate the largest force in the mode with the highest elastic modulus. Molecular interactions between polymer chains in the system likely help to diffuse this stress concentration. In a saturated system, where the interactions are interrupted, an increase in stress concentration would cause areas of high stress to fail, with the failure propagating through the polymer along a plane perpendicular to the applied stress.

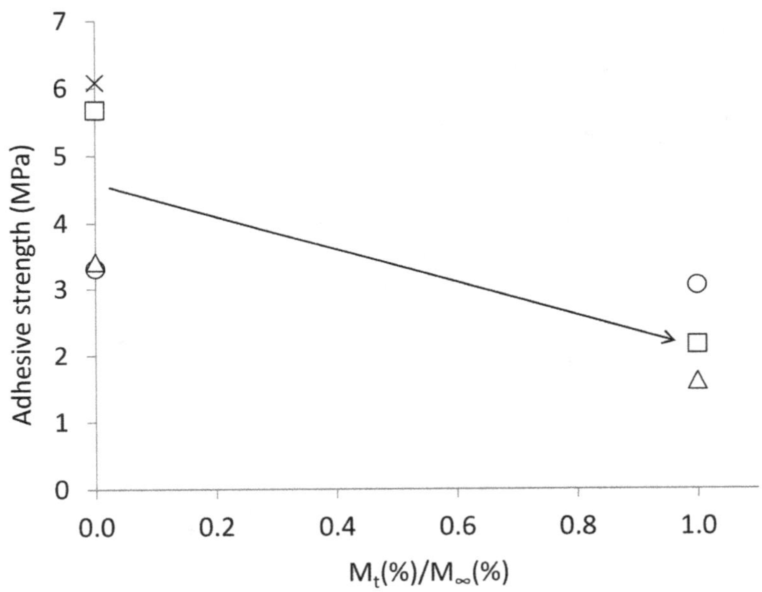

Figure 16. Adhesive strength vs. water content [10].

8. States of water absorbed in coating

As it is understood that water molecules absorbed into the polymer network interrupt the intermolecular bonding, which causes degradation of mechanical properties, we attempt to estimate the manner in which absorbed water bonds to the polymer network [11]. That is, we characterize the absorbed water into states based on the results of DSC and attenuated total reflection (ATR) FT-IR spectroscopy.

Higuchi and Liu have reported that there are three ways the absorbed water exists within polymer [74–76]: one is as free water, which is free of any forces and whose freezing point is normally around 0°C; another is as weakly bound freezable water in which the water freezing point shifts to lower temperature due to intermolecular interaction between water molecules and the polymer network. In our research, the temperature was between -40°C and -65°C, and the other is as strongly bound nonfreezable water, where the water molecules do not freeze at any temperature due to their strong bonds with the polymer network. The type and mass of each water molecule can be determined by observing the position and area of the exothermic peak in the DSC chart during the cooling process.

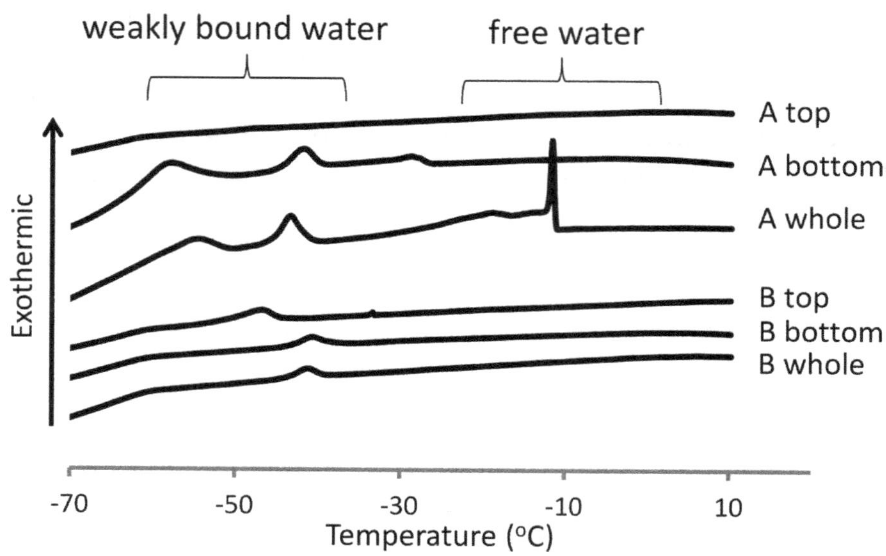

Figure 17. DSC curves for coating A and B during cooling process [11].

	Percentage (%)			
	Free	Weak 1	Weak 2	Strong
A top	0	0.3	0	99.7
A bottom	0	7.9	12.1	80.0
A whole	5.9	5.7	6.3	82.1
B top	0	8	0	92.0
B bottom	0	2.6	0	97.4
B whole	0	3.6	0	96.1

Table 1. Percentage of water in each state estimated by DSC when polymer is saturated with water [11].

Figure 17 shows the DSC curves for coatings A and B, which are the same as those in section 7. As seen in **Figure 17**, the free water is observed in A whole which is around −10°C. The peaks due to weakly bound freezable water are observed in A bottom, A whole, B top, B bottom, and B whole at around between −40°C and −65°C. Free water is not observed in coating B. **Table 1** shows numerically the percentage of water in each state estimated by DSC when polymer is saturated. Calculated values obtained from DSC measurements are compared with a gravimetric measurement of the total mass of absorbed water. The mass of water stored in each observed DSC state is determined, and the remaining water is assumed to be stored in a strongly bound state. As seen in **Table 1**, most of the water molecules, above 80%, are strongly bound to the polymer network. The rest are weakly bound. Free water is present only between layers.

As for FT-IR analysis, Musto [77], Mijovic [78], and Cotugno [79] have reported that there are three types of water molecules, depending on the relative strength of intermolecular bonding to the polymer network, resulting in a difference in the wavelength of stretching vibration of O–H bonds of water molecules, S_0, S_1, and S_3. S_2 represents two hydrogen atoms participating in hydrogen bonding to polymer network. S_1 represents only one hydrogen atom participating in hydrogen bonding. S_0 represents that there are no hydrogen atoms participating in hydrogen bonding. S_2 can be detected at lower frequency due to strong hydrogen bonding, while S_0 can be at higher frequency due to weak hydrogen bonding. **Figure 18** shows a 2D correlation spectroscopy of FT-IR for the A top layer, which can determine the time-resolved correlation during water absorption between any two points of the spectrum. **Figure 18(a)** shows a synchronous correlation, which presents a correlation between changes in different peaks. **Figure 18(b)** shows an asynchronous correlation, which presents a phase difference between changes in different peaks. As seen in the synchronous plots in **Figure 18(a)**, absorbed water creates the largest peak at around 3460/cm^{-1}, which correlates to OH stretching vibration. There are two noticeable shoulders on the primary water peak, which indicates additional peaks centered at around 3260 and 3600/cm^{-1}. Based on the consideration mentioned above, these peaks correspond to S_1, S_2, and S_0 states at 3460, 3260, and 3600/cm^{-1}. In the asynchronous plots in **Figure 18(b)**, the S_2 shoulder shows a phase delay in its growth behind the S_1 peak. There is no significant asynchronous correlation between S_1 and S_0, which suggests that S_0 and S_1 form first and then S_2.

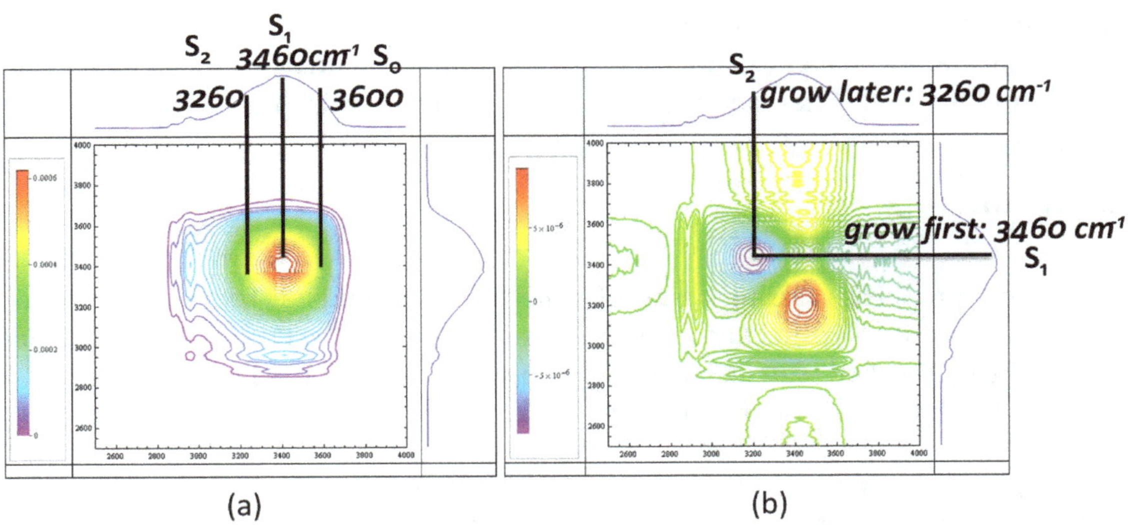

Figure 18. 2D correlation spectroscopy of FT-IR for A top layer. (a) Synchronous plot and (b) asynchronous plot [11].

From a comparison of areas of FT-IR peaks of S_0, S_1, and S_2 of A top, A bottom, B top, and B bottom, the relative strengths are $S_1 > S_2 > S_0$ for urethane and $S_2 > S_1 > S_0$ for epoxy. This indicates that epoxy has more hydrophilic functional groups that present potential bonding sites for water molecule than urethane. Epoxy presents the groups for water, namely, tertiary and secondary amine, hydroxyl, and ether groups, while urethane presents the groups for water, namely, secondary amine, ether, and carbonyl groups.

Taken together, the DSC and FT-IR data indicate that six possible active functional groups are present in the epoxy/urethane coating system. Depending on the intensity, according to Harazaki [80], amine, hydroxyl, ether, carbonyl, and isocyanate groups provide sites for a strong DSC state due to hydrogen bonding. Epoxide groups provide sites for weak 2 due to dipole-dipole force. Other groups provide site for weak 1 due to van der Waals force. In addition, we speculate that all functional groups produce sites for the S_1 or S_2 state in FT-IR.

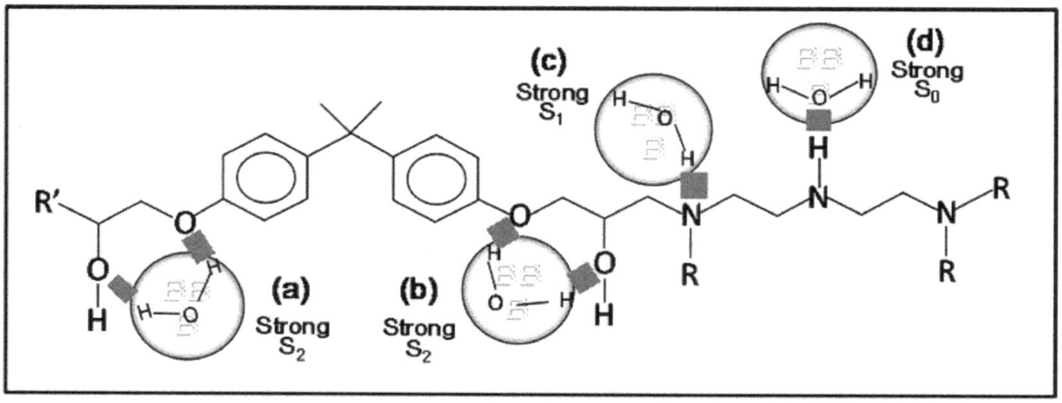

Figure 19. Image of water molecule absorbed in polymer network. (a) and (b) present strong DSC and S_2 FT-IR states, (c) presents strong DSC and S_1 FT IR states, and (d) presents strong DSC and S_0 FT IR states

Finally, an image of water molecules bonding to the polymer network is shown in **Figure 19**. In this image, water molecules (a) and (b) present strong DSC and S_2 FT-IR states, (c) presents strong DSC and S_1 FT-IR states, and (d) presents strong DSC and S_0 FT-IR states.

9. Conclusion

The rheological properties during aging of waterborne coatings at room temperature and during degradation in an artificially accelerated environment have been investigated, and the manner in which absorbed water bonds to the polymer network has been emphasized as an aggressive factor that causes degradation.

Viscoelastic properties represent that E′ and E″ vary intensively depending on the process of aging and degradation. During aging, E′ generally increases with time, while E″ decreases. In the process of degradation, both E′ and E″ decrease accordingly. The viscoelastic property determines whether it presents solid-like or liquid-like response from the comparison of relative strengths of the relaxation time (τ) and operating time (t). E′, E″, and tanδ can be used to evaluate likely planes of failure in multilayer combinations.

The majority of the water molecules are strongly bound to the polymer network through hydrogen bonds with polar functional groups, which destroys intermolecular bonding between macromolecules and reduces the bulk materials ability to diffuse stress concentrations and thereby lowers a coating's overall strength.

Acknowledgements

The authors thank Dr. T. Handa and R. Nishio of NTT-AT and T. Kamisho of NTT East for their helpful discussions and Richard Jackson and Ethan Becker, interns from the University of Wisconsin Platteville, for their enthusiastic work in this research.

Author details

Yukitoshi Takeshita*, Takashi Miwa, Azusa Ishii and Takashi Sawada

*Address all correspondence to: takeshita.yukitoshi@lab.ntt.co.jp

NTT Device Innovation Center, NTT Corporation, Atsugi-shi, Kanagawa, Japan

References

[1] Takeshita Y, Xiaoxi Z, Sugiyama A, Sawada T: Increasing the lifetime and reliability of telecommunication plant materials. NTT Technical Review. 2010; Available from: https://www.ntt-review.jp/archive/ntttechnical.php?contents=ntr201301fa4_s.html [Accessed: 2015-05-01]

[2] NTT East Technical Support Center. 4 Material degradation. In: Trouble Q & A in Telecommunication Facilities. 2nd ed. The Telecommunication Association: Ohmsha; 2008. p. 309–415

[3] Takeshita Y, A study of rheological property for degradation of environmentally friendly coating. In :Proceedings of the Reports on Weathering Technology Achievements 2010; 25 November 2010; JWTC: Tokyo; 2010. p. 25–34

[4] Takeshita Y. Sakata S. Sawada T. Jackson Jr R. Nishio R. Rheological property and adhesive performance of polymer coating film for telecommunication plant. Corr. Eng. 2011; 60: 88–95. Allerton Press, Inc., 2011

[5] Takeshita Y. Jackson Jr R. Sakata S. Sawada T. Nishio R. Aging property of eco-friendly anticorrosion organic coating film for telecommunication plant. Polym. Eng. Sci. 2012; 52:572–580. doi: 10.1002/pen.22119

[6] Kamisho T. Takeshita Y. Sakata S. Sawada T. Study on performance evaluation of heavy-duty anticorrosive coating for telecommunication facilities by dynamic viscoelasticity. Bosei-Kanri. 2011; 55:449–454.

[7] Kamisho T. Takeshita Y. Sakata S. Sawada T. Aging behaviour of water-based and solvent-based coatings in dry and wet environments. Bosei-Kanri. 2013; 57:451–457.

[8] Kawai H, editor. Handbook of Industrial Coating. 1st ed. Techno System: Tokyo; 2008. 256 p.

[9] Kamisho T. Takeshita Y. Sakata S. Sawada T. Water absorption of water-based anticorrosive coatings and its effect on mechanical property and adhesive performance. J. Coat. Tech. Res. 2014; 11:199–205. doi: 10.1007/s11998-013-9517-z

[10] Takeshita Y. Becker E. Sakata S. Kamisho T. Miwa T. Sawada T. Water absorption and degraded stress relaxation behaviour in water-borne anticorrosive urethane/epoxy coatings. J. Chem. Chem. Eng. 2015; 9:75–89. doi: 10.17265/1934-7375/2015.02.001

[11] Takeshita Y., Becker E., Sakata S., Miwa T., Sawada T. States of water absorbed in water-borne urethane/epoxy coatings. Polymer. 2014; 55:2505–2513. doi: 10.1016/j.polymer.2014.03.027

[12] Perera Y. D. Physical ageing of organic coatings. Prog. Org. Coat. 2003; 47:61–76. doi: 10.1016/S0300-9440(03)00037-7

[13] Sell C. G., McKenna G. B. Influence of physical ageing on the yield response of model DGEBA/poly(propylene oxide) epoxy glasses. Polymer. 1992; 33:2103–2113. doi: 10.1016/0032-3861(92)90876-X

[14] Hutchinson M. J. Physical aging of polymers. Prog. Polym. Sci. 1995; 20:703–760. doi: 10.1016/0079-6700(94)00001-I

[15] Mijovic J., Devine S. T., Ho T. Physical aging in poly(methyl methacrylate)/poly(styrene-co-acrylonitrile) blends. I. Stress relaxation. J. Appl. Polym. Sci. 1990; 39:1133–1151. doi: 10.1002/app.1990.070390509

[16] Ho T., Mijovic J., Lee C. Effect of structure on stress relaxation of polymer blends in glassy state. Polymer. 1991; 32, 619–627. doi:10.1016/00323861(91)90473V

[17] Mijovic J., Ho T., Kwei K. T. Physical aging in poly(methyl methacrylate)poly(styrene-co-acrylonitrile) blends. Part II: Enthalpy relaxation. Polym. Eng. Sci. 1989; 29:1604–1610. doi: 10.1002/pen.760292210

[18] Tavares C. A., Gulmine V. J., Lepienski M. C., Akcelrud L. The effect of accelerated aging on the surface mechanical properties of polyethylene. Polym. Degrad. Stab. 2003; 81,367–373. doi: 10.1016/S0141-3910(03)00108-3

[19] Gulmine V. J., Janissek R. P., Heise M. H., Akcelrud L. Degradation profile of polyethylene after artificial accelerated weathering. Polym. Degrad. Stab. 2003; 79:385–397. doi: 10.1016/S0141-3910(02)00338-5

[20] Cook D. W., Mehrabi M., Edward H. G. Ageing and yielding in model epoxy thermosets. Polymer. 1999; 40:1209–1218. doi: 10.1016/S0032-3861(98)00343-7

[21] Sell G. C., McKenna B. G. Influence of physical aging on the yield behavior of model DGEBA/poly(propylene oxide) epoxy glasses. Polymer. 1992; 33:2103–2113. doi: 10.1016/0032-3861(92)90876-X

[22] Kong S. E. Physical aging in epoxy matrices and composites. in Advances in Polymer Science 80, Springer-Verlag: Berlin Heidelberg;1986. 125 p. doi: 10.1007/3–540-16423-5_14

[23] Davis J. W., Pethrick A. R. Investigation of physical ageing in polymethylmethacrylate using positron annihilation, dielectric relaxation and dynamic mechanical thermal analysis. Polymer 1998; 39: 255–266. doi: 10.1016/S0032-3861(97)00392-3

[24] Robertson G. C., Wilkes L. G., Physical aging behavior of miscible blends of poly(methyl methacrylate) and poly(styrene-co-acrylonitrile). Polymer. 2001; 43,1581–1589. doi: 10.1016/S0032-3861(00)00556-5

[25] Fechine J. M. G., Rabello S. M., Souto-Maior M. R., Catalani H. L. Surface characterization of photodegraded poly(ethylene terephthalate). The effect of ultraviolet absorbers. Polymer. 2004; 45:2303–2308. doi: 10.1016/j.polymer.2004.02.003

[26] Fechine J. M. G., Rabello S. M., Souto-Maior M. R. The effect of ultraviolet stabilizers on the photodegradation of poly(ethylene terephthalate). Polym. Degrad. Stab. 2002; 75:153–159. doi: 10.1016/S0141-3910(01)00214-2

[27] Cowie M. G. J., Harris S., McEwen J. I. J. Physical ageing in poly(vinyl acetate) 1. Enthalpy relaxation. Polym. Sci. B. 1997; 35:1107–1116. doi: 10.1002/(SICI)1099-0488(199705)35:7 <1107::AID-POLB9>3.0.CO;2-S

[28] Linde van der R. Belder G. E., Perera Y. D. Effect of physical aging and thermal stress on the behavior of polyester/TGIC powder coatings. Prog. Org. Coat. 2000; 40:215–224. doi: 10.1016/S0300-9440(00)00125-9

[29] Perera Y. D., Schutyser P. Effect of physical aging on thermal stress development in powder coatings. Prog. Org. Coat. 1994; 24.299–307. doi: 10.1016/0033-0655(94)85021-6

[30] Perera Y. D., Eynde V. D. Moisture and temperature inducted stresses (hygrothermal stresses) in organic coatings. J. Coat. Technol. 1987; 59:55–63.

[31] Skaja A., Fernando D., Croll S. Mechanical property changes and degradation during accelerated weathering of polyester-urethane coatings. JCT Res.. 2006; 3:41–45. doi: 10.1007/s11998-006-0004-7

[32] Perera Y. D. Effect of thermal and hygroscopic history on physical ageing of organic coatings. Prog. Org. Coat. 2002; 44: 55–62. doi: 10.1016/S0300-9440(01)00241-7

[33] Grentzer H. T., Holsworth M. R., Provder T. The application of the dynamic mechanical analyzer to organic coatings. ACS Org. Coat. Plastics. 1981; 44: 515–519

[34] Perera Y. D., Schutyser P. Effect of physical aging on properties of organic coatings. In: Proceedings of the 22nd FATIPEC Congress vol. 1; 1994; Budapest. p. 25–38

[35] Skrovanek J. D., Schoff K. C., Thermal mechanical analysis of organic coatings. Prog. Org. Coat. 1988; 16:135–163. doi: 10.1016/0033-0655(88)80011-6

[36] Chang P. E. Viscoelastic windows of pressure-sensitive adhesives. J. Adh. 1991; 34:189–200. doi: 10.1080/00218469108026513

[37] Kaelble H. D. Peel adhesion: Influence of surface energies and adhesive rheology. J. Adh.. 1969; 1: 102–123. doi: 10.1080/00218466908078882

[38] Gent N. A., Petrich P. R. Adhesion of viscoelastic materials to rigid substrates. Proc. Roy. Soc. A. 1969; 310:433–448. doi: 10.1098/rspa.1969.0085

[39] Chan H. K., Howard J. G., Structure-property relationships in acrylic adhesives. J. Adh. 1978; 9:279–304. doi:10.1080/00218467808075121

[40] Gent N. A., Hamed R. G. Peel mechanics. J. Adh. 1975; 2:91–95. doi: 10.1080/00218467508075041

[41] Kraus G., Rollmann W. K., Gray A. R. Tack and viscoelasticity of block copolymer based adhesives. J. Adh. 1979; 10:221–236. doi: 10.1080/00218467908544626

[42] Class B. J., Chu G. S. The viscoelastic properties of rubber–resin blends. I. The effect of resin structure. J. Appl. Polym. Sci. 1985; 30:805–814. doi: 10.1002/app.1985.070300229

[43] Class B. J., Chu G. S. The viscoelastic properties of rubber–resin blends. II. The effect of resin molecular weight. J. Appl. Polym. Sci. 1985; 30:815–824. doi: 1002/app. 1985.070300230

[44] Class B. J., Chu G. S. The viscoelastic properties of rubber–resin blends. III. The effect of resin concentration. J. Appl. Polym. Sci. 1985; 30:825–842. doi: 10.1002/app. 1985.070300231

[45] Chu G. S. in Chapter 8, Viscoelastic properties of pressure-sensitive adhesives, D. Satas, editor. Handbook of Pressure Sensitive Adhesive Technology. Van Nostrand Reinhold: New York (1989).

[46] Takemura A., Kuriyama H., Moriguchi T., Tomita B., Mizumachi H., Dynamic mechanical properties and adhesive strengths of polyacrylates I. Mokuzai Gakkaishi. 1986; 32:813–819 (in Japanese)

[47] Ochi K., Iwakoshi M., Niiho M. Properties of cured epoxide resins (Part 2). J. Adh. Soc. Jpn. 1974; 10:266–272 (in Japanese).

[48] Motohashi K., Toota B., Mizumachi H., Sakaguchi H. Temperature dependency of bond strength of polyvinyl acetate emulsion adhesives for wood. Wood Fiber Sci. 1984; 16:72–85.

[49] Mizumachi H., Hatano Y., Yamagishi Y., Keon L. Influence of the mechanical properties of plasticized polymers on wood adhesion. Holzforschung. 1983; 37:153–156. doi: 10.1515/hfsg.1983.37.3.153

[50] Mizumachi H., Tsukiji M. Interaction between wood and polymers. Pt. VI. Dynamic mechanical properties of polymers filled with wood components. Hozforschung. 1980; 34:122–124. doi: 10.1515/hfsg.1980.34.4.122

[51] Kobayashi M., Hatano Y., Tomita B. Viscoelastic properties of liquefied wood/epoxy resin and its bond strength. Holzforschung. 2001; 55:667–671. doi: 10.1515/HF.2001.108

[52] Mizumachi H., Fujino M. Interaction between wood and polymers. Holzforschung. 1972; 26:164–169. doi: 10.1515/hfsg.1972.26.5.164

[53] Takemura A., Tomita B., Mizumachi H., Dynamic mechanical properties and adhesive strengths of epoxy resins modified with liquid rubber. I. Modification with ATBN. J. Appl. Polym. Sci. 1985; 30:4031–4043. doi: 10.1002/app.1985.070301007

[54] Takemura A., Shiozawa K., Tomita B., Mizumachi H. Dynamic mechanical properties and adhesive strengths of epoxy resins modified with liquid rubber. II. Modification with CTBN. J. Appl. Polym. Sci. 1986; 31:1351–1362. doi: 10.1002/app.1986.070310519

[55] Davies R. O., Jones G. O. The irreversible approach to equilibrium in glasses. Proc. Roy. Soc. A.1953; 217:26–42. doi: 10.1098/rspa.1953.0044

[56] Davies R. O., Jones G. O. Thermodynamic and kinetic properties of glasses. Adv. Phys. 1953; 2:370–410. doi: 10.1080/00018735300101252

[57] The Society of Rheology, Japan. What's rheology? Available from: http://www.srj.or.jp/index-j.html [Accessed: 2016-05-02]

[58] Crank J. 9.4. Sorption- and desorption-time curves. 10.6.7 Use of initial rates of sorption and desorption. In The Mathematics of Diffusion. 2nd ed. Oxford University Press; Oxford; 1975. p. 179–189 and 244–246. ISBN-10: 0198534116

[59] Ferry D. J. In Viscoelastic Properties of Polymers. 3rd ed. John Wiley & Sons: New York; 1980. ISBN: 978–0-471–04894-7

[60] Williams G., Watts C. D. Non-symmetrical dielectric relaxation behaviour arising from a simple empirical decay function. Trans. Faraday. Soc. 1970; 60:80–85. doi: 10.1039/TF9706600080

[61] Hetman P., Szabat B., Weron K., Wodziński D. 2003. On the rajagopal relaxation-time distribution and its relationship to the Kohlrausch–Williams–Watts relaxation function. J. Non-Cryst. Solids. 2003; 330:66–74. doi: 10.1016/j.jnoncrysol.2003.08.060

[62] Johnston D. Stretched exponential relaxation arising from a continuous sum of exponential decays. Phys. Rev. B. 2008; 74:184430–1-184430–7. doi: http://dx.doi.org/10.1103/PhysRevB.74.184430

[63] Lindsey, C. P., Patterson G. D. Detailed comparison of the Williams–Watts and Cole–Davidson functions. J. Chem. Phys. 1980; 73:3348–3357. doi: http://dx.doi.org/10.1063/1.440530

[64] Weron K. Relaxation in glassy materials from lévy stable distributions. Acta Physica Polonica A. 1986; 70:529–539.

[65] Barton J. M. Effect of absorbed water on the thermal relaxation of biaxially stretched crosslinked poly(methyl methacrylate. Polymer. 1979; 20:1018–1024. doi: 10.1016/0032-3861(79)90202-7

[66] Baschek G. Hartwig G. Zahradnik F. Effect of water absorption in polymers at low and high temperatures. Polymer. 1999; 40:3433–3441. doi: 10.1016/S0032-3861(98)00560-6

[67] Yoon S. C., Ratner B. D. Surface and bulk structure of segmented poly(ether urethanes) with perfluoro chain extenders. 2. FTIR, DSC, and X-ray photoelectron spectroscopic studies. Macromolecules.1988; 21:2392–2400. doi: 10.1021/ma00186a016

[68] Funke W. Zorll U. Murthy B. G. Interfacial effects in solid paint films related to some film properties. J Paint Technol. 1969; 41: 210–221.

[69] Sperling L. H. Introduction to Physical Polymer Science. 4th ed. Wiley; Hoboken, New Jersey; 2005. ISBN: 978-0-471-70606-9

[70] Bahar I., Erman B., Fytas G., Steffen W. Intramolecular contributions to stretched-exponential relaxation behavior in polymers. Macromolecules.1994; 27:5200–5205. doi: 10.1021/ma00096a051

[71] Ping Z. H., Nguyen Q. T., Chen S. M., Zhou J. Q., Ding Y. D. States of water in different hydrophilic polymers — DSC and FTIR studies. Polymer. 2001; 42:8461–8467. doi: 10.1016/S0032-3861(01)00358-5

[72] Rosa F., Bordado J., Casquilho M. Kinetics of water absorbency in AA/AMPS copolymers: applications of a diffusion–relaxation model. Polymer. 2002; 43:63–70. doi: 10.1016/S0032-3861(01)00596-1

[73] Nogueira P., Ramarez C., Torres A., Abad M. J., Cano J., Lopez J. Effect of water sorption on the structure and mechanical properties of an epoxy resin system. J. Appl. Polym. Sci. 2001; 80:71–80. doi: 10.1002/1097-4628(20010404)80:1<71::AID-APP1077>3.0.CO;2-H

[74] Higuchi A., Iijima T. D.S.C. investigation of the states of water in poly(vinyl alcohol) membranes. Polymer. 1985; 26: 1207–1211. doi: 10.1016/0032-3861(85)90254-X

[75] Higuchi A., Iijima T. D.S.C. investigation of the states of water in poly(vinyl alcohol-co-itaconic acid) membranes. Polymer. 1985; 26: 1833–1837. doi: 10.1016/0032-3861(85)90011-4

[76] Liu W. G., Yao K. D. What causes the unfrozen water in polymers: hydrogen bonds between water and polymer chains? Polymer. 2001; 42:3943–3947. doi:10.1016/S0032-3861(00)00726-6

[77] Musto P., Ragosta G., Mascia L. Vibrational spectroscopy evidence for the dual nature of water sorbed into epoxy resins. Chem. Mater. 2000; 12;1331–1341. doi: 10.1021/cm9906809

[78] Mijovic J., Zhang H. Local dynamics and molecular origin of polymer network-water interactions as studied by broadband dielectric relaxation spectroscopy, FTIR, and molecular simulations. Macromolecules. 2003; 36:1279-1288. doi: 10.1021/ma021568q

[79] Cotugno S., Mensitieri G., Musto P., Sanguigno L. Molecular interactions in and transport properties of densely crosslinked networks: a time-resolved FT-IR spectroscopy investigation of the epoxy/H_2O system. Macromolecules. 2005; 38:801–811. doi: 10.1021/ma040008j

[80] Harazaki Y. Fundamental Science of Coatings. 1st ed. Maki Press: Tokyo; 1977. 32 p.

Permissions

All chapters in this book were first published in SCT, by InTech Open; hereby published with permission under the Creative Commons Attribution License or equivalent. Every chapter published in this book has been scrutinized by our experts. Their significance has been extensively debated. The topics covered herein carry significant findings which will fuel the growth of the discipline. They may even be implemented as practical applications or may be referred to as a beginning point for another development.

The contributors of this book come from diverse backgrounds, making this book a truly international effort. This book will bring forth new frontiers with its revolutionizing research information and detailed analysis of the nascent developments around the world.

We would like to thank all the contributing authors for lending their expertise to make the book truly unique. They have played a crucial role in the development of this book. Without their invaluable contributions this book wouldn't have been possible. They have made vital efforts to compile up to date information on the varied aspects of this subject to make this book a valuable addition to the collection of many professionals and students.

This book was conceptualized with the vision of imparting up-to-date information and advanced data in this field. To ensure the same, a matchless editorial board was set up. Every individual on the board went through rigorous rounds of assessment to prove their worth. After which they invested a large part of their time researching and compiling the most relevant data for our readers.

The editorial board has been involved in producing this book since its inception. They have spent rigorous hours researching and exploring the diverse topics which have resulted in the successful publishing of this book. They have passed on their knowledge of decades through this book. To expedite this challenging task, the publisher supported the team at every step. A small team of assistant editors was also appointed to further simplify the editing procedure and attain best results for the readers.

Apart from the editorial board, the designing team has also invested a significant amount of their time in understanding the subject and creating the most relevant covers. They scrutinized every image to scout for the most suitable representation of the subject and create an appropriate cover for the book.

The publishing team has been an ardent support to the editorial, designing and production team. Their endless efforts to recruit the best for this project, has resulted in the accomplishment of this book. They are a veteran in the field of academics and their pool of knowledge is as vast as their experience in printing. Their expertise and guidance has proved useful at every step. Their uncompromising quality standards have made this book an exceptional effort. Their encouragement from time to time has been an inspiration for everyone.

The publisher and the editorial board hope that this book will prove to be a valuable piece of knowledge for researchers, students, practitioners and scholars across the globe.

List of Contributors

Daniel Picchi, Mateus Lourenço, Alexandre da Silva, Daniel Nascimento Jr., Eric Saldanha, Inácio Dantas and José Resende
Elektro Eletricidade e Serviços S.A. (Elektro Electricity Services), Campinas, Brazil

Ghada Al-Hudhud
Department of Information Technology, King Saud University, Riyadh, Saudi Arabia

Benchagra Mohamed
University Hassan, ENSA, Khouribga, Morocco

Mircea Georgescu and Daniela Popescul
"Alexandru Ioan Cuza" University, Iaşi, Romania

Danilo Hernane Spatti
Federal Technological University of Paraná, Curitiba, Brazil

Luisa Helena Bartocci Liboni
Federal Institute of Education, Science and Technology of São Paulo, University of São Paulo, São Paulo, Brazil

Ivana Semanjski and Sidharta Gautama
Ghent University, Department of Telecommunications and Information Processing, Ghent, Belgium

Joseph S.T. Smalley, Felipe Vallini, Abdelkrim El Amili and Yeshaiahu Fainman
Department of Electrical & Computer Engineering, University of California, San Diego, La Jolla, CA, USA

Abdelfatteh Haidine, Sanae El Hassani, Abdelhak Aqqal and Asmaa El Hannani
Information Technologies Laboratory (LTI), National School of Applied Sciences, El Jadida, Morocco

John Jefferson Antunes Saldanha, Eduardo Machado dos Santos, and Ana Paula Carboni de Mello
Federal University of Pampa, Alegrete, Rio Grande do Sul, Brazil

Daniel Pinheiro Bernardon
Federal University of Santa Maria, Santa Maria, Rio Grande do Sul, Brazil

Yukitoshi Takeshita, Takashi Miwa, Azusa Ishii and Takashi Sawada
NTT Device Innovation Center, NTT Corporation, Atsugi-shi, Kanagawa, Japan

Index

www.ingramcontent.com/pod-product-compliance
Lightning Source LLC
LaVergne TN
LVHW080707210926
848485LV00003B/430